"Information through Innovation"

THE CONCEPTS OF DATABASE MANAGEMENT

PHILIP J. PRATT
Grand Valley State University

bf
boyd & fraser publishing company

Acquisitions Editor: Anne E. Hamilton
Production Manager: Peggy J. Flanagan
Director of Production: Becky Herrington
Compositor: Huntington & Black Typography
Interior Design: Becky Herrington
Cover Design: Diana Coe
Cover Photo: Melanie Carr
Manufacturing Coordinator: Tracy Megison

© 1995 by boyd & fraser publishing company
One Corporate Place • Ferncroft Village
Danvers, MA 01923

International Thomson Publishing
boyd & fraser publishing company is an ITP company.
The ITP trademark is used under license.

This book is printed on recycled, acid-free paper that meets
Environmental Protection Agency standards.

Manufactured in the United States of America

Library of Congress Cataloging-in-Publication Data

```
Pratt, Philip J., 1945
   The concepts of database management/ Philip J. Pratt,
   p.  cm.
   Includes index.
   ISBN 0-87709-779-8
   1. Database management.  I. Title.
QA76.9.D3P729287  1994
005,75'6-dc20                           9410049
                                          CIP
```

1 2 3 4 5 6 7 8 9 10 KE 8 7 6 5 4

To My Loving Family

Preface xi
About This Book xi
Special Features xii
Organization of the Textbook xiii
Acknowledgements xv

CHAPTER 1   Introduction to Database Management 1

1.1   INTRODUCTION 1
1.2   BACKGROUND 10
1.3   ADVANTAGES OF DATABASE PROCESSING 15
1.4   DISADVANTAGES OF DATABASE PROCESSING 17
      SUMMARY 18
      KEY TERMS 19
      EXERCISES 19

CHAPTER 2   Data Models 20

2.1   INTRODUCTION 20
2.2   PREMIERE PRODUCTS 21
2.3   THE RELATIONAL MODEL 25
2.4   THE NETWORK MODEL 30
2.5   THE HIERARCHICAL MODEL 34
      SUMMARY 36
      KEY TERMS 37
      EXERCISES 38

CHAPTER 3   The Relational Model: Data Definition and Manipulation 39

3.1   INTRODUCTION 39
3.2   SQL 40
3.3   QBE 51
3.4   THE RELATIONAL ALGEBRA 55
3.5   NATURAL LANGUAGES 57
      SUMMARY 59
      KEY TERMS 59
      EXERCISES 59

CHAPTER 4   Relational Model II: Advanced Topics 61

4.1   INTRODUCTION 61
4.2   VIEWS 61
4.3   INDEXES 64
4.4   THE CATALOG 67
4.5   INTEGRITY RULES 68
4.6   CHANGING THE STRUCTURE OF A RELATIONAL DATABASE 69
4.7   WHAT DOES IT TAKE TO BE RELATIONAL? 71
      SUMMARY 72
      KEY TERMS 73
      EXERCISES 73

CHAPTER 5   Database Design I: Normalization 74

5.1   INTRODUCTION 74
5.2   FUNCTIONAL DEPENDENCE 75
5.3   KEYS 77
5.4   FIRST NORMAL FORM 78
5.5   SECOND NORMAL FORM 79
5.6   THIRD NORMAL FORM 82
5.7   INCORRECT DECOMPOSITIONS 85
      SUMMARY 89
      KEY TERMS 89
      EXERCISES 89

CHAPTER 6   Database Design II: Design Methodology 91

6.1   INTRODUCTION 91
6.2   INFORMATION-LEVEL DESIGN 92
6.3   THE METHODOLOGY 92
6.4   DATABASE DESIGN EXAMPLES 97
6.5   PHYSICAL-LEVEL DESIGN 108
      SUMMARY 109
      KEY TERMS 110
      EXERCISES 110

CHAPTER 7   Functions of a Database Management System 113

7.1   INTRODUCTION 113
7.2   STORAGE AND RETRIEVAL 114
7.3   CATALOG 114
7.4   SHARED UPDATE 115
7.5   RECOVERY 123
7.6   SECURITY 124
7.7   INTEGRITY 125
7.8   DATA INDEPENDENCE 127
7.9   UTILITIES 129
      SUMMARY 130
      KEY TERMS 130
      EXERCISES 130

CHAPTER 8   Database Administration 132

8.1   INTRODUCTION 132
8.2   POLICY FORMULATION AND IMPLEMENTATION 133
8.3   DATA DICTIONARY MANAGEMENT 138
8.4   TRAINING 138
8.5   DBMS SUPPORT 138
8.6   DATABASE DESIGN 142
      SUMMARY 142
      KEY TERMS 143
      EXERCISES 143

CHAPTER 9   Application Generation 144

9.1   INTRODUCTION 144
9.2   APPLICATIONS 145
9.3   APPLICATION GENERATORS: COMPONENTS 149
9.4   APPLICATION GENERATORS: UTILIZATION 155

9.5 APPLICATION GENERATORS: OTHER TERMINOLOGY 157
    SUMMARY 157
    KEY TERMS 158
    EXERCISES 158

Answers to Odd-Numbered Exercises 161

CHAPTER 1 – INTRODUCTION TO DATABASE MANAGEMENT 161
CHAPTER 2 – DATA MODELS 162
CHAPTER 3 – THE RELATIONAL MODEL: DATA DEFINITION AND MANIPULATION .. 162
CHAPTER 4 – RELATIONAL MODEL II: ADVANCED TOPICS 164
CHAPTER 5 – DATABASE DESIGN I: NORMALIZATION 165
CHAPTER 6 – DATABASE DESIGN II: DESIGN METHODOLOGY 166
CHAPTER 7 – FUNCTIONS OF A DATABASE MANAGEMENT SYSTEM 167
CHAPTER 8 – DATABASE ADMINISTRATION 168
CHAPTER 9 – APPLICATION GENERATION 169

GLOSSARY 171
BIBLIOGRAPHY 176
INDEX 177

Database courses in leading computer science programs have been offered comprehensively since the 1970s to students aspiring to careers in data processing. In their professions, such students often become involved in the design, development, implementation, and maintenance of large, mainframe, database-oriented application systems. Some become involved in related areas such as database administration.

Until recently, such professionals were the only segment of the population that had any direct contact with databases and database management systems. With the advent of microcomputer database management systems, however, the picture has changed dramatically. Virtually all segments of the population can now be considered potential users of such systems, including such diverse groups as home computer owners, owners of small businesses, and end users in large organizations. Where recently the spreadsheet was the tool that most microcomputer owners felt was essential, many are now turning to database management systems as the essential tool.

The major microcomputer database systems have continually added features to increase the ease with which they may be used so users can begin to apply the products relatively quickly. *Truly* effective operation of such a product, however, requires more than just knowledge of the product itself, although that is obviously important. It requires a general knowledge of the database area, including such topics as database design, database administration, and application development using these systems. Although the depth of understanding required is certainly not as great for the majority of users as it is for the data processing professional, a lack of any understanding in these areas precludes effective use of the product in all but the most limited applications.

ABOUT THIS BOOK

This book is intended for anyone who is interested in becoming familiar with database management. It is appropriate for students in introductory database classes in computer science or information systems programs. It is also appropriate for students in database courses in related disciplines such as business at either the undergraduate or graduate level. Such students require a general understanding of the database environment. In addition, *The Concepts of Database Management* is ideal for courses introducing students of any discipline to database management, which is becoming increasingly more popular. It is also appropriate for individuals considering the purchase of a microcomputer database package who want to make effective use of such a package.

This book assumes that students have some familiarity with personal computers. A single introductory course is all the background that is required. Although students need not have any background in programming to effectively use this text, those with a little programming background will be able to explore certain topics in more depth than students without such a background.

Although database management on mainframes is discussed in the book, the main thrust is on microcomputer database management. For a more thorough discussion of database management on mainframes, see [9], [10], [11], [13], [14], or [15] in the bibliography at the end of this book.

SPECIAL FEATURES

Emphasis on the Relational Model

The main emphasis of this book is on the relational model. Not only is the relational model becoming the dominant model for mainframe systems, but microcomputer systems are almost exclusively relational.

Coverage of the Other Models

Although the other models, specifically the hierarchical and network models, are typically not found on microcomputer systems, it is important that students be exposed to these models. The other models shed light on alternative approaches to a problem. A brief study of them also gives students a better perspective on the relational model, its relationship to other approaches, and its place in database management. Thus, these models are presented in the book, but only at a very intuitive level.

Data Manipulation within the Relational Model

The most important approach to data manipulation in the years to come is clearly SQL. This language has now been adopted as a standard. Many systems already support it, and most of those that do not, will probably add such support in the not-too-distant future. Thus, gaining familiarity with SQL is crucial. This book contains not only a discussion of SQL, but also many examples and exercises through which students can gain the needed familiarity.

There are also other approaches to data manipulation within the model. Perhaps the three most important are Query-By-Example (QBE), the relational algebra, and natural languages, all of which are covered in this book.

Database Design

The important process of database design is given detailed treatment. A methodology is presented that represents a highly useful subset of the full-scale database design methodology given in the bibliography [14]. For most microcomputer users, the design methodology as presented in this book will be more than enough to allow them to develop a correct design for whatever requirements they face. Those few who become involved in the design of a major mainframe database or an especially complex microcomputer database can use the bibliography [14] to quickly and easily step into the material on database design.

Functions Provided by a DBMS

Current microcomputer database management systems offer such a wide variety of features, that students must know the functions such systems *should* provide. These functions are presented and discussed.

Database Administration (DBA)

Although the office of database administration (DBA) is absolutely essential in the mainframe environment, it is also important in a microcomputer environment, especially if the database is to be shared among several users. Thus, this book includes a detailed discussion of the database administration function.

DBMS Selection

The process of selecting a DBMS is important, given the myriad systems that are available. Unfortunately, it is not an easy task. To help students make such selections effectively, the book includes a detailed discussion of the process and offers a comprehensive checklist.

Applications Generators
The book includes a detailed discussion of what an application system is, what an applications generator is, and how an applications generator is used in the creation of an application system. The relationships between the terms *applications generator*, *fourth-generation language*, and *fourth-generation environment* is presented as part of the discussion of applications generators. The dBASE applications generator is illustrated as a specific example of this type of tool.

Glossary
The glossary contains definitions of the important terms in the book.

Numerous Realistic Examples
The book contains numerous examples illustrating each of the concepts. The examples are realistic and representative of the kinds of problems that are encountered in the design, manipulation, and administration of databases.

Exercises
The book contains a wide variety of questions. At key points within the chapters, the students are asked questions to ensure they understand the material before they proceed. The answers to these questions are given immediately following the questions. At the end of each chapter, there are exercises that test the students' recall of the important points in the chapter and their ability to apply what they have learned. The answers to the odd-numbered exercises are given at the end of the book.

Instructor's Manual
The accompanying Instructor's Manual contains detailed teaching tips, answers to exercises in the book, test questions (and answers), and transparency masters.

ORGANIZATION OF THE TEXTBOOK

The textbook consists of nine chapters that deal with general database topics and are not geared to any specific database management system. A brief description of the organization of topics in the chapters follows.

Introduction
Chapter 1 provides a general introduction to the field of database management.

Data Models
Chapter 2 presents the concept of a data model. The relational, network, and hierarchical models are covered at an intuitive level.

The Relational Model
The relational model is covered in detail in Chapters 3 and 4. Chapter 3 covers the data definition and manipulation aspects of the model using SQL, QBE, the relational algebra, and natural languages. Chapter 4 covers some advanced aspects of the model such as views, the use of indexes, the catalog, and the relational integrity rules. It also includes a discussion of the question: *What does it take to be relational?*

Database Design

Chapters 5 and 6 are devoted to database design. Chapter 5 covers the normalization process, which provides a mechanism to find and correct bad designs. In Chapter 6, a methodology for database design is presented and illustrated through a number of examples.

Functions of a Database Management System

Chapter 7 discusses the features that should be provided by a full-functioned microcomputer database management system.

Database Administration

Chapter 8 is devoted to the role of database administration. Also included in this chapter is a discussion of the process of selecting a DBMS.

Applications Generators

Chapter 9 deals with the process of applications generation. It covers applications generators, their features, and their use in the process. It also describes the term *fourth-generation* and how it relates to the term *applications generator*.

ACKNOWLEDGEMENTS

I would like to acknowledge the individuals who made contributions in the preparation of this book.

I appreciate the efforts of the following individuals who reviewed the text and gave many helpful suggestions: Judy Adamski, Jenison, Michigan; Cindy L. Bonfini-Hotlosz, West Virginia Northern Community College; Linda Denny, Sinclair Community College; Mary Last, Grand Valley State University; John M. Lloyd, Montgomery County Community College; Michael Michaelson, Palomar College; Robert E. Norton, San Diego Mesa College; Hung-Lian Tang, Bowling Green State University; and Francis Whittle, Dutchess Community College.

The efforts of the following members of the staff of boyd & fraser publishing company have been invaluable: Thomas K. Walker, President and CEO; James H. Edwards, Executive Editor; Anne Hamilton, Acquisitions Editor; Peggy Flanagan, Production Manager; and Becky Herrington, Director of Production. I would also like to express my appreciation to Ginny Harvey for her many helpful suggestions.

Introduction to Database Management

OBJECTIVES

1. Provide a general introduction to the field of database management.
2. Introduce some basic terminology.
3. Describe the advantages and disadvantages of database processing.

1.1 INTRODUCTION

Chris, Henry, Pat, and Maxine are all taking a microcomputer database course at a local college. Each of the four is interested in learning how to use a microcomputer to manage data effectively, and each one has come to the class with a special project in mind.

Case 1: Chris Chris is the director of volunteers for a large, service-oriented organization. When volunteers are needed for a project, it's his responsibility to locate and enlist them. Currently, he keeps the information he needs on cards in a filing cabinet. For each volunteer, he has a card that contains pertinent information. In a separate drawer, he has a card for each project in which his organization is involved. Some of his responsibilities, such as determining the skills of a particular volunteer, involve locating a single card. This aspect of his job is not very complicated. Other responsibilities, such as gathering the names of all the volunteers who have a particular skill, are much more difficult, since they involve examining every card in the cabinet. Preparing reports is also a time-consuming process.

What Chris would really like is an easy way to enter all this information into his computer and to get it back when he needs it. In particular, he would like to be able to ask the computer such questions as the following:

- What interests and/or skills are required for a given project?
- Which volunteers have a particular combination of skills and have not worked on any projects in the last two years?

He also would like to be able to group volunteers from the same family and ask the following question:

- Which families have more than one person who is suited to work on a given project?

Case 2: Henry   Henry owns a chain of four bookstores. Henry has used a computerized file-oriented system to organize the data he uses to run his bookstores.

He gathers and organizes information about publishers, authors, and books. Each book has a code that uniquely identifies the book. In addition, he records the title, the publisher, the type of book, the price, and whether the book is paperback. He also records the author or authors of the books along with the number of units of the book that are in stock in each of the branches.

Henry uses this information in a variety of ways. For example, a customer may be interested in books written by a certain author or of a certain type. Henry wants to be able to tell the customer which books by the author or of that type he currently has in stock. If the customer wants a book that is not in stock at one branch, Henry needs to be able to determine if any of the other branches currently have it.

Case 3: Pat   Pat is the office manager for a dental practice. She has been asked by one of the dentists in the practice to look into acquiring a computer to manage all of the appointment scheduling, billing, and preparation of insurance forms. This is a complex task that requires a wide range of information. The following summary of terms used in the dental practice illustrates the extent of the information that is needed.

There are two types of *providers* of services: *dentists* and *hygienists*. *Patients* have *appointments*. Each appointment is *scheduled* with a given provider on a given *date* at a given *time* in a given *room* and for some specific collection of *services*. Patients are grouped into *households* (families), which are ultimately responsible for paying for any services rendered. Patients in a given household may all be covered by the same dental *insurance* policy or they may be covered by different policies. Because patients have this insurance through their *employers*, it is important to relate patients to employers and employers to the insurance companies whose policies cover their employees. Also, if more than one patient in a household is covered by an insurance policy, it is important to distinguish between *primary* and *secondary insurance carriers*. Within a household, a given insurance policy may be the primary carrier for one patient and the secondary carrier for another patient.

Case 4: Maxine   Maxine schedules sections of courses at a college. For the purpose of simplicity, we will call these sections *classes*. She determines which faculty member will teach which class and when, and assigns each class to a room. Each room has certain characteristics, such as the number of seats, the types of lab and video display equipment present, and so on. Each class has its own set of requirements for a room in which it may be held. Each faculty member has certain courses that he or she is capable of teaching as well as specific courses that he or she would prefer to teach. In making her schedules, Maxine needs to match rooms, classes, and faculty in such a way that there are no conflicts, while making sure that faculty members are capable of teaching the classes to which they are assigned. In addition, she tries to assign faculty members to the courses that they prefer to teach.

With regard to the microcomputer database course they are taking, what is it that Chris, Henry, Pat, and Maxine all have in common? They all need to be able to store and retrieve data in an efficient and organized way. Further, all of them are interested in more than one category of information. In database management we call these categories **entities**.

Chris is interested in such entities as volunteers, projects, and households. Henry is interested in books, authors, publishers, and branches. Pat is interested in patients, providers (dentists and hygienists), households, services, appointments, insurance companies, rooms, and so on. Maxine is interested in classes, faculty members, rooms, and times.

Besides wanting to store data that pertains to more than one entity, the four students are also interested in relationships between the entities. Chris, for example, is interested not only in volunteers, projects, and households but also in the relationship between volunteers and projects (which volunteers are assigned to which projects), between volunteers and households (which volunteers are a part of which households), and so on. Henry is interested in relating books to the authors who wrote them, to the publishers who published them, and to the branches that have them in stock.

What these students most want to know is how to maintain and use a database. A **database** is a structure that contains information about many kinds of entities and about the relationships between the entities. Henry's database, for example, would contain information about books, authors, branches, and publishers. It would provide facts that related authors to the books they directed and branches to the books they currently have in stock. With the use of a database, Henry would be able to start with a particular book and find out who wrote it as well as which branches have it. Alternatively, he could start with an author and find all the books he or she wrote, together with the publishers of these books.

Approach 1: Writing Programs to Maintain a Database

One possible approach to the problem would be to use a high-level language like BASIC to write programs. Some of the programs would allow users to enter and modify data. Others would be used to produce reports. Still others might be used to calculate statistics about the data that has been stored. Some programs could produce special documents. To handle customer billing, a collection of programs would include a program that printed invoices. Such a collection for payroll would have to contain programs to print paychecks and W-2 forms. Other programs could be devised to answer such questions as, "Do we currently have available any volunteers with skills X, Y, and Z?"

Such a collection of programs is often called a **system of programs** or a **software system**; sometimes it is simply called a **system**. Commercial developers often call these systems of programs **software packages**. They may also be called an **application system** or an **application package**. The programs that make up the collection are often called **application programs**.

Case 2: Henry and BASIC   Using BASIC, Henry had created such a system to maintain and report on books, publishers, authors, and branches several years ago. The system that Henry created used several **files**. (If you have written programs before, you are familiar with the concept of a file. You may have referred to this type of file as a **data file**. For those of you who are not familiar with this concept, some sample files, represented in the form of tables, are shown in Figure 1.1a. The rows in the table are called **records**, and the columns are called **fields**.)

BRANCH

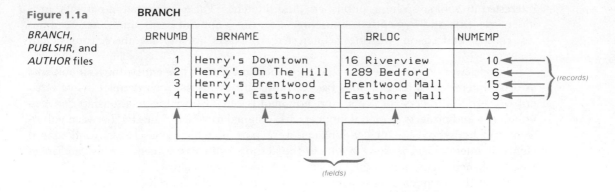

(records)

(fields)

PUBLSHR

PUBCODE	PUBNAME	PUBCITY
AH	Arkham House Publ.	Sauk City, Wisconsin
AP	Arcade Publishing	New York
AW	Addison-Wesley	Reading, Mass.
BB	Bantam Books	New York
BF	Boyd and Fraser	Boston
JT	Jeremy P. Tarcher	Los Angeles
MP	McPherson and Co.	Kingston
PB	Pocket Books	New York
RH	Random House	New York
RZ	Rizzoli	New York
SB	Schoken Books	New York
SI	Signet	New York
TH	Thames and Hudson	New York
WN	W.W. Norton and Co.	New York

AUTHOR

AUTHNUMB	AUTHNAME
1	Archer, Jeffrey
2	Christie, Agatha
3	Clarke, Arthur C.
4	Francis, Dick
5	Cussler, Clive
6	King, Stephen
7	Pratt, Philip
8	Adamski, Joseph
10	Harmon, Willis
11	Rheingold, Howard
12	Owen, Barbara
13	Williams, Peter
14	Kafka, Franz
15	Novalis
16	Lovecraft, H. P.
17	Paz, Octavio
18	Camus, Albert
19	Castleman, Riva
20	Zinbardo, Philip
21	Gimferrer, Pere
22	Southworth, Rod
23	Wray, Robert

Figure 1.1b

Description of
BRANCH,
PUBLSHR, and
AUTHOR files

BRANCH

NAME	DESCRIPTION
BRNUMB	Branch number
BRNAME	Branch name
BRLOC	Branch location
NUMEMP	Number of employees

PUBLSHR

NAME	DESCRIPTION
PUBCODE	Publisher code
PUBNAME	Publisher name
PUBCITY	Publisher city

AUTHOR

NAME	DESCRIPTION
AUTHNUMB	Author number
AUTHNAME	Author name

The first three of Henry's files are shown in Figure 1.1a. Descriptions of the fields are given in 1.1b. The files are called *BRANCH*, *PUBLSHR*, and *AUTHOR*.

(**Note:** There are many cases where names are limited to eight characters. To make sure that we would not have problems in such cases, names in this book will adhere to the eight-character limit. This is why the name for the publisher file has been shortened to *PUBLSHR*.)

The *BOOK* file is shown in Figure 1.2a with a description of its fields appearing in Figure 1.2b on the next page.

Figure 1.2a

BOOK file

BOOK

BK CODE	BKTITLE	PUB CODE	BK TYPE	BK PRICE	PB
0180	Shyness	BB	PSY	7.65	T
0189	Kane and Abel	PB	FIC	5.55	T
0200	The Stranger	BB	FIC	8.75	T
0378	The Dunwich Horror and Others	PB	HOR	19.75	F
079X	Smoke-screen	PB	MYS	4.55	T
0808	Knockdown	PB	MYS	4.75	T
1351	Cujo	SI	HOR	6.65	T
1382	Marcel Duchamp	PB	ART	11.25	T
138X	Death on the Nile	BB	MYS	3.95	T
2226	Ghost from the Grand Banks	BB	SFI	19.95	F
2281	Prints of the 20th Century	PB	ART	13.25	T
2766	The Prodigal Daughter	PB	FIC	5.45	T
2908	Hymns to the Night	BB	POE	6.75	T
3350	Higher Creativity	PB	PSY	9.75	T
3743	First Among Equals	PB	FIC	3.95	T
3906	Vortex	BB	SUS	5.45	T
5163	The Organ	SI	MUS	16.95	T
5790	Database Systems	BF	CS	54.95	F
6128	Evil Under the Sun	PB	MYS	4.45	T
6328	Vixen 07	BB	SUS	5.55	T
669X	A Guide to SQL	BF	CS	23.95	T
6908	DOS Essentials	BF	CS	20.50	T
7405	Night Probe	BB	SUS	5.65	T
7443	Carrie	SI	HOR	6.75	T
7559	Risk	PB	MYS	3.95	T
7947	dBASE Programming	BF	CS	39.90	T
8092	Magritte	SI	ART	21.95	F
8720	The Castle	BB	FIC	12.15	T
9611	Amerika	BB	FIC	10.95	T

Figure 1.2b

Description of
BOOK file

Q & A

BOOK

NAME	DESCRIPTION
BKCODE	Book code
BKTITLE	Book title
PUBCODE	Book publisher code
BKTYPE	Book type
BKPRICE	Price
PB	Paperback? (T or F)

Question:

To check your understanding of the relationship between publishers and books, answer the following questions: Who published *Knockdown*? Which books did Signet publish?

Answer:

The publisher code (*PUBCODE*) in the row in the *BOOKS* table for *Knockdown* is PB. Examining the *PUBLSHR* table, we see that PB is the code assigned to Pocket Books.

To find the books published by Signet, we look up its code in the *PUBLSHR* table and see that it is SI. Next, we look for all records in the *BOOK* table for which the publisher code is SI and find that Signet published *Cujo, Carrie, The Organ,* and *Magritte.*

Figure 1.3a

WROTE and
INVENT files

WROTE

BK CODE	AUTH NUMB	SEQ NUMB
0180	20	1
0189	1	1
0200	18	1
0378	16	1
079X	4	1
0808	4	1
1351	6	1
1382	17	1
138X	2	1
2226	3	1
2281	19	1
2766	1	1
2908	15	1
3350	10	1
3350	11	2
3743	1	1
3906	5	1
5163	12	2
5163	13	1
5790	7	1
5790	8	2
6128	2	1
6328	5	1
669X	7	1
6908	22	1
7405	5	1
7443	6	1
7559	4	1
7947	7	1
7947	23	2
8092	21	1
8720	14	1
9611	14	1

INVENT

BK CODE	BR NUMB	OH
0180	1	2
0189	2	2
0200	1	1
0200	2	3
079X	2	1
079X	3	2
079X	4	3
1351	1	1
1351	2	4
1351	3	2
138X	2	3
2226	1	3
2226	3	2
2226	4	1
2281	4	3
2766	3	2
2908	1	3
2908	4	1
3350	1	2
3906	2	1
3906	3	2
5163	1	1
5790	4	2
6128	2	4
6128	3	3
6328	2	2
669X	1	1
6908	2	2
7405	3	2
7559	2	2
7947	2	2
8092	3	1
8720	1	3
9611	1	2

The table called *WROTE* in Figure 1.3a is used to relate books and authors. The sequence number indicates the order in which the authors of a particular text should be listed. The table called *INVENT* in the same figure is used to indicate the number of units of a particular book that are currently on hand at a particular branch. The first row, for example, indicates that there are two units of the book whose code is 0180 currently on hand at Branch 1. Descriptions of these tables are shown in Figure 1.3b.

Figure 1.3b

Description of
WROTE and
INVENT files

WROTE	
NAME	DESCRIPTION
BKCODE	Book code
AUTHNUMB	Author number
SEQNUMB	Sequence number. Used to indicate order in which multiple authors should be listed.

INVENT	
NAME	DESCRIPTION
BKCODE	Book code
BRNUMB	Branch number
OH	Number of units on hand

Q & A

Question:

To check your understanding of the relationship between authors and books, answer the following questions: Who wrote *The Organ*? (Be sure to list the authors in the correct order.) Which books did Jeffrey Archer write?

Answer:

To determine who wrote *The Organ*, we first examine the *BOOK* table to find its book code (5163). Next we look for all rows in the *WROTE* table in which the book code (*BKCODE*) is 5163. There are two such rows. In one of them the author number (*AUTHNUMB*) is 12, and in the other, it is 13. All that is left is to look in the *AUTHOR* table to find the authors who have been assigned the numbers 12 and 13. The answer is Barbara Owen (12) and Peter Williams (13). The sequence number for author 12 is 2, however, and the sequence number for author 13 is 1. Thus, listing the authors in the proper order, the authors are Peter Williams and Barbara Owen.

To find the books written by Jeffrey Archer, we look up his number in the *AUTHOR* table and find that it is 1. Then we look for all rows in the *WROTE* table for which the author number is 1. There are three such rows. The corresponding book codes are 0189, 2766, and 3743. Looking up these codes in the *BOOK* table, we find that Jeffrey Archer wrote *Kane and Abel*, *The Prodigal Daughter*, and *First Among Equals*.

Question:

A customer in Branch 1 wishes to purchase *The Vortex*. Is it currently in stock in Branch 1?

Answer:

Looking up the code for *The Vortex* in the *BOOK* table, we find it is 3906. To find out how many copies are in stock in Branch 1, we look for a row in the *INVENT* table with 3906 in the *BKCODE* column and 1 in the *BRNUMB* column. Since there is no such row, Branch 1 doesn't have any copies of *The Vortex*.

Question:

We would like to obtain a copy of *The Vortex* for this customer. Which other branches currently have it in stock and how many copies do they have?

Answer:

We already know that the code for *The Vortex* is 3906. (If we didn't, we would simply look it up in the *BOOK* table.) To find out the branches that currently have copies, we look for rows in the *INVENT* table with 3906 in the *BKCODE* column. There are two such rows. The first one indicates that Branch 2 currently has one copy. The second indicates that Branch 3 currently has two copies.

Henry wrote several programs to allow him to update the data in his files. Separate programs were required to allow for the addition, correction, and deletion of books, publishers, and authors. Henry also had to create programs to produce any of the reports he wanted. Further, the logic that we just discussed for relating books and publishers and for relating authors and books had to be built into these programs.

**Approach 2:
Using a
Database
Management
System**

In the system that Henry developed in BASIC, he effectively implemented a database. He certainly maintained data on several entities (books, publishers, and authors). He also maintained relationships between these entities. To the computer, however, Henry was dealing with nothing more than a collection of isolated files. It was only through the efforts of Henry's programs that the crucial relationships were maintained. Another way of stating this would be to say that Henry's programs *managed the database*. As Henry discovered, this can be a *very complex task*.

Fortunately, we no longer necessarily have to create our own programs, because the computer is now able to assist in managing the database for us. The tool it uses is called a **database management system**, or **DBMS**. A DBMS is a program or collection of programs whose function is to manage a database on behalf of the people who use it. It greatly simplifies the task of manipulating and using a database. In fact, had Henry used a DBMS, he might not have had to write *a single program*! (*Note:* You will have a chance to see exactly how you can use one of the most popular DBMSs to manage a database in the module that follows this chapter and in subsequent modules.)

Case 1: Chris and a DBMS   Chris decided that a microcomputer DBMS could fulfill the needs of his service organization. He determined the structure of the database he needed (this is called **designing a database**), following the procedure discussed in the course that he took. His database contained the information he needed about volunteers, projects, and households. He then communicated this design to the DBMS and began to enter data. He found that through the use of the DBMS, all of the reports he needed were easy to produce. Since then, he has had no need to resort to any programming.

Case 2: Henry and a DBMS   Even though he already had the system he developed in BASIC, Henry was so impressed with what he learned about databases and database management systems that he decided to purchase a DBMS. He then designed his database and communicated the design to the DBMS. He entered his data and was easily able to produce a number of reports. He could now perform many important tasks without writing any programs at all.

However, the DBMS did not completely satisfy Henry's needs. He found, for example, that when he had entered PA as the publisher code for a particular book, the DBMS simply let him do it, *even though there was no publisher with code PA on file*. In the programs that he had written in BASIC, he had included a feature to prevent this sort of thing from happening. He also discovered that some of his most important reports could not be produced with the built-in facilities of the DBMS.

**Approach 3:
Programming
with a DBMS**

Fortunately, many DBMSs allow users to write programs to supplement the built-in features of the DBMS. Some allow programs to be written in existing languages such as BASIC, Pascal, or COBOL. These languages are expanded to include additional commands that allow for the accessing and updating of a database. Other DBMSs include their own language. These languages usually contain all of the typical structures one expects in a modern programming language, along with a number of special features that are geared toward manipulation of the database. Many DBMSs also offer features that provide screen and printer management. The languages furnished by the better DBMSs provide features that are superior to those found in many non-DBMS languages (for example, BASIC, Pascal, COBOL, and FORTRAN).

Case 2: Henry and a DBMS   Henry was pleased to discover that his DBMS included its own programming language. He found that with the DBMS he could write programs to overcome the deficiencies he had spotted. Using a program, he could ensure that a book could not be entered if the book's publisher was not already in the database. This programming capability enabled him to produce reports that were beyond the basic capabilities of the DBMS.

Henry was also pleasantly surprised to find that he could develop programs with the DBMS much more rapidly than he had developed the BASIC programs in his previous system. Simple commands could be used in his programs to accomplish various important tasks, such as accessing the database, interacting with the screen, printing lines on reports, copying data from one file to another, and so on. He found tools within the system to allow the development of large portions of some of the important programs. He is convinced that he has now found the perfect environment for his system.

Approach 4: Using a Commercial Software Package

Case 3: Pat, a Software Package, and a DBMS   For a variety of reasons, Pat decided not to use a DBMS herself to develop the system for the dental practice. First, the many entities and complicated relationships made the requirements so complex that she felt the problem should be handled by a computer professional. Pat had another reason for not wanting to undertake the task. She was very leery of attempting to develop a system on which the whole billing operation of the dental practice would depend. She believed there were many issues involved in the development and use of such a system that she didn't understand, and that seemed to be yet another reason for putting such an undertaking in the hands of professionals.

That's the direction that the dental practice took. They purchased a software package from an organization that specialized in software for dental offices, and it seems to have fit their needs quite well. The package is geared specifically toward appointment scheduling, billing, and the preparation of insurance forms.

Pat did decide she could do certain things with a DBMS, however. In fact, even though the dental practice purchased the software package, she determined that her office had some needs that more than justified the purchase of a DBMS. She has a number of things in mind to do with it. She plans to track the history of various types of treatment by household and by family. She is also going to track the use of supplies in various types of treatment. The package that the practice purchased does not do these things, and she feels comfortable doing them herself with the DBMS. The purposes for which she uses the DBMS will be completely separate from those of the package that the practice purchased.

Approach 5: Solving Part of the Problem with a DBMS

Case 4: Maxine and a DBMS   Maxine originally intended to purchase a software package to handle her scheduling needs. In her estimation, however, none of the packages she looked at were really suitable. Some would require her to change the way she carried out the scheduling process; others did not supply certain reports that she felt were essential. At this point, she happened to be taking a database course and wondered if she might be able to develop the system herself with a DBMS.

Maxine talked to the professor who was teaching the course. In the process, she discovered that simply using a DBMS would not solve her problem. The DBMS could be used to maintain crucial information about rooms, faculty, and classes, but the actual process of scheduling was beyond its built-in capabilities.

This meant that someone would have to write programs to do the scheduling. The scheduling process itself is very intricate, and the task of writing programs to implement it seemed complicated as well. Maxine felt she was not in a position to undertake such a formidable task. Happily, the professor decided to have a team of students try to develop the system Maxine needed as their project for a course they would enroll in during the next semester.

The professor also pointed out to Maxine that she could use the DBMS to carry out certain productive activities even before the full system was developed. Even if she made out the schedule entirely by hand without the benefit of the computer, she could still derive benefits from the DBMS. She could begin by manually entering her schedule into a database. Although that might sound like extra work, once it had been done, she could easily get a variety of reports that would be very useful. She could print out the schedule of each faculty member and the schedule of all the classes in a given room, or the times and locations at which the various sections of a given course were offered — that is, a time schedule. If someone needed to schedule a room for a special offering, such as a seminar, she could have the system search for rooms that would be open at the desired time, would be capable of seating the number of people expected to attend, and would contain any special equipment that was required.

Maxine is excited both by the possibility of the students at the college developing the complete system she needs and by the amount of useful work she can accomplish even before this has been done. She is eager to begin.

The interests and needs of these four students are typical of those of today's microcomputer users. The foregoing approaches to their problems are all being used by people around the world today. Ever increasing numbers of people are finding database management systems the ideal tool for solving a wide variety of problems.

Section 1.2 presents some background material on database management and some of the most commonly used terms. In sections 1.3 and 1.4, we will examine some of the advantages and disadvantages of database management systems.

1.2 BACKGROUND

This section introduces some terminology and concepts that are very important in the database environment. Some of the terms will be familiar to you from the material in the preceding section.

Entities, Attributes, and Relationships

The most fundamental terms are entity, attribute, and relationship. An **entity** is really just like a noun; it is a person, place, or thing. The entities of interest to Henry, for example, are such things as publishers, branches, and books. The entities that are of interest to Chris include members and projects. In her dental office, Pat is interested in such entities as patients, households, services, providers of services, appointments, and so on.

An **attribute** is a property of an entity. The term is used here exactly as it is used in everyday English. For the entity *person*, for example, the list of attributes might include such things as eye color and height. For Henry, the attributes of interest for the entity *book* are such things as code, title, type of book, price, and so on.

Figure 1.4 shows two entities, *PUBLSHR* and *BOOK*. It also shows a number of attributes. The *PUBLSHR* entity has three attributes: publisher code (*PUBCODE*), publisher name (*PUBNAME*), and publisher city (*PUBCITY*). The attributes are really just the columns in the table. The *BOOK* entity has six attributes: book code (*BKCODE*), book title (*BKTITLE*), publisher code (*PUBCODE*), book type (*BKTYPE*), book price (*BKPRICE*), and paperback (*PB*). (The last attribute simply indicates whether or not the book is paperback.)

The final key term is relationship. When we speak of a **relationship**, we really mean an association between entities. There is an association between publishers and books, for example. A publisher is associated with all of the books that it publishes, and a book is associated with its publisher. Technically, we say that a publisher is *related to* all of the books it publishes, and a book is *related to* its publisher.

This particular relationship is called a **one-to-many** relationship. *One* publisher is associated with *many* books, but each book is associated with only *one* publisher. (In this type of relationship, the word *many* is used differently than in everyday English; it may not always literally mean a large number. In this context, it would mean that a publisher can be associated with *any number* of books.)

A one-to-many relationship is often represented pictorially in the fashion shown in Figure 1.5. In such a diagram, entities and attributes are represented in precisely the same way as they are shown in Figure 1.4. The relationship is represented by an arrow. The "one" part of the relationship, in this case *PUBLSHR*, is indicated by a single-headed arrow, and the "many" part of the relationship, in this case *BOOK*, is indicated by a double-headed arrow.

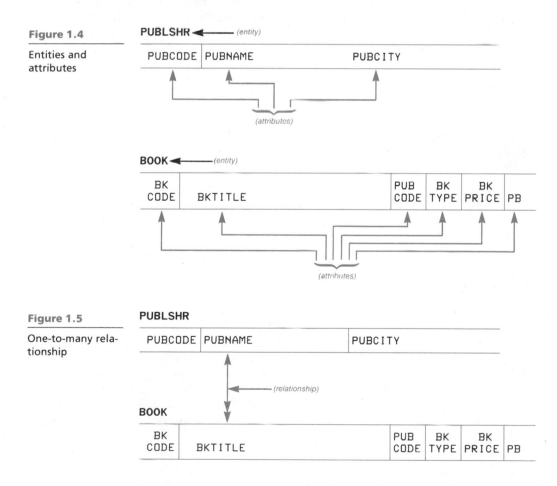

Figure 1.4

Entities and attributes

Figure 1.5

One-to-many relationship

Files and Databases

You encountered the word *file* earlier in this chapter. If you have done some programming yourself, you are probably already familiar with the word. Basically, a file used to store data, which is often called a *data file*, is the computer counterpart to an ordinary paper file you might keep in a filing cabinet. You will recall that Chris kept such a file on the members in his organization. His filing cabinet was filled with cards, one for each member. The crucial aspect of this type of file is that it houses information on a *single entity* and the attributes of that entity. In Chris's case, the single entity was *member*. Each card kept information on the crucial attributes of one member of the organization.

A database is much more than a file, however. Unlike a typical data file, a database can house information about more than one entity. And there is another difference. A database also holds information about the relationships among the various entities. Not only would Henry's database have information about both books and publishers, for example, it would also hold information relating publishers to the books they had produced. Formally, the definition of a database is as follows:

Definition: A **database** is a structure that can house information about multiple types of entities, the attributes of these entities, and the relationships among the entities.

Database Management Systems

Managing a database is inherently a complicated task. Fortunately, software packages called **database management systems** can do the job of manipulating databases for us. A database management system, or **DBMS**, is a software product through which users interact with a database. The actual manipulation of the underlying database is handled by the DBMS. In some cases, users may interact with the DBMS directly, as shown in Figure 1.6a. In other cases, users may interact with programs; these programs in turn interact with the DBMS (see Figure 1.6b). In either case, it is only the DBMS that actually accesses the database.

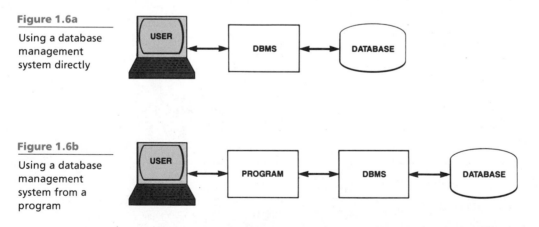

Figure 1.6a

Using a database management system directly

Figure 1.6b

Using a database management system from a program

Using a DBMS, for example, Henry can request the system to find publisher PB and the system will either locate this publisher and give us the data or tell us that no such publisher exists in the database. All the work involved in this task is performed by the DBMS. If publisher PB is in the database, Henry can then ask for the books this publisher has published and again the system will perform all the work involved in locating these books. Likewise, when Henry stores a new book in the database, the DBMS performs all the tasks necessary to ensure that the book is related to the appropriate publisher.

Mainframe DBMSs have been in use since the 1960s. They have continually been enhanced over the years, gaining in selection of features and in performance. Recently, microcomputer DBMSs that possess many of the features of their mainframe counterparts have become available. The leaders in this field, such as dBASE™ from Ashton-Tate and R:BASE® from Microrim, are also improved on a continuous basis. They make the power of database management available to large numbers of microcomputer users. The focus of this text is microcomputer database management systems. For a discussion of database management on mainframes, see [9], [10], [11], [13], [14], and [15] in the references at the end of the book.

Database Processing

When we use the term **database processing**, we mean that the data to be processed is stored in a database and the data in the database is being manipulated by a DBMS. We have seen how database processing will benefit Chris, Henry, Pat, and Maxine with their individual systems. Still greater benefit is obtained by combining the activities of several users and allowing them to share a common database.

Let's first consider the nondatabase approach illustrated in Figure 1.7. Mary, Jeff, and Joan are three separate users at the same college. Mary is involved in enrolling students in courses, in producing class lists, and so on. She has her own system of programs and files that she uses to perform this activity on the computer. Her files contain information on classes, on faculty members who teach these classes, and on the students who are enrolled in them.

Figure 1.7

Nondatabase approach

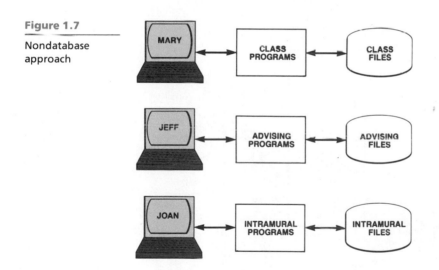

Jeff is involved in the advising process, that is, the process of advising students about their programs and their progress toward a degree. He has his own system of programs and files. His files, which are totally separate from Mary's, contain information on faculty members, on the students who are advised by them, and on the requirements that have already been fulfilled by these students.

Joan is in charge of maintaining information on the intramural athletic program. She too has her own system of programs and files. Her files contain information on the various sports that are available, the teams that participate in these sports, the students who belong to these teams, and the current records of the teams.

Two major problems arise with this nondatabase approach. The first problem is duplication of data. Mary, Jeff, and Joan are each keeping information about students, for example. Presumably, each of them will need the address of all the students. Thus the address of each student is stored in at least three separate places in the computer. Not only is this wasteful of space, but it causes a real headache when a student moves and his or her address must be changed.

The second problem is that it is extremely difficult to fulfill requirements that involve data from more than one system. The format of the files might not even be compatible from one system to another. The following question and answer segment demonstrates such a requirement.

Q & A

Question: Suppose we want to list the number, name, and address of a particular student. We also want to list the classes in which the student is currently enrolled, the name of the student's advisor, and the intramural sports in which the student is participating. Where would we find the necessary data?

Answer: The student's number, name, and address could come from any one of the three systems. The classes in which the student is enrolled could be found in Mary's system. The name of the student's advisor would be found in Jeff's system. The sports in which the student is participating could be found in Joan's system. Thus, this requirement involves data from all three systems.

By contrast, in a database approach, instead of having separate collections of files, Mary, Jeff, and Joan would be able to share a common database managed by a DBMS (see Figure 1.8).

Figure 1.8

Database approach (using a database management system)

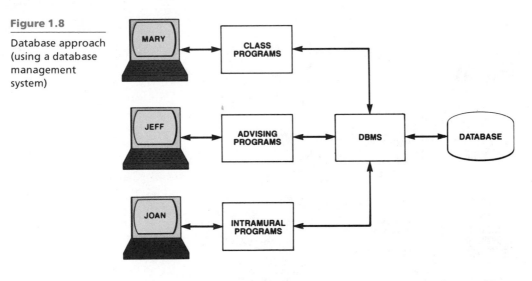

Each student would appear only once, so his or her address would likewise appear only once. No space would be wasted, and changing a student's address would be a very simple procedure. Further, since all the data would be in a single database, listing the information on a student, the student's classes, the student's advisor, and the sports in which the student was participating would now be quite possible. In fact, with a good DBMS, it should be a simple task.

1.3 ADVANTAGES OF DATABASE PROCESSING

The database approach to processing, using a microcomputer DBMS, offers ten clear advantages. They are listed in Figure 1.9 and are discussed below.

Figure 1.9

Advantages of database processing

Advantages of Database Processing

1. Lower Cost
2. Getting More Information from the Same Amount of Data
3. Sharing of Data
4. Balancing Conflicting Requirements
5. Controlled or Eliminated Redundancy
6. Consistency
7. Integrity
8. Security
9. Increased Productivity
10. Data Independence

1. **Lower cost**. This is an interesting advantage. In discussions about mainframe DBMSs, cost is usually listed as a disadvantage (see [14]). By the time all the appropriate components have been purchased, the total price can easily run between $100,000 and $400,000. The size and complexity of a DBMS may also necessitate the use of more hardware. Purchasing (or leasing) this additional hardware represents another cost.

   On microcomputers, the opposite is true. Some of the more limited (but still very useful) systems are priced under $100. Even for the most powerful and most sophisticated of these microcomputer DBMSs, the price ranges between $500 and $800. These prices make the features of the DBMS available to a wide range of users.

2. **Getting more information from the same amount of data**. The primary goal of a computer system is to turn data (recorded facts) into information (the knowledge gained by processing these facts). In a nondatabase environment, data is often partitioned into several disjointed systems, each system having its own collection of files. Any request for information that would necessitate accessing data from more than one of these collections can be extremely difficult. In many cases, for all practical purposes, it is considered impossible. Thus, the desired information is unavailable, not because it is not stored in the computer, but because of the way it has been broken down into the various collections of files. When, instead, all the data for the various systems is stored in a single database, the information becomes available. Given the power of a modern DBMS, not only is the information available, but the process of getting it can be a quick and easy one.

3. **Sharing of data**. The data of various users can be combined and shared among authorized users, allowing all users access to a greater pool of data. Several users can have access to the same piece of data, for example, a customer's address, and still use it in a variety of ways. When an address is changed, however, the new address immediately becomes available to all users. In addition, new applications can be developed through the use of the existing data in the database without the burden of having to create separate collections of files.

4. **Balancing conflicting requirements.** For the database approach to function adequately within an organization, a person or group should be in charge of the database itself, especially if it is to serve a number of users. This person or group is often called **Database Administration (DBA)**. By keeping the overall needs of the organization in mind, DBA can structure the database in such a way that it benefits the entire organization, not just a single group. While this may mean that an individual user group is served less well than it would have been if it had its own isolated system, the organization as a whole is better off. And ultimately, when the organization benefits, so do the individual groups of users.

5. **Controlled or eliminated redundancy.** With database processing, data that was formerly kept separate in a nondatabase system is integrated into a single database, so multiple copies of the same data no longer exist. With the nondatabase approach, Mary, Jeff, and Joan each had a copy of the address of each student, but with the database approach, each student's address would occur only once, thus eliminating the duplication (technically called **redundancy**).

   Eliminating redundancy not only saves space but makes the process of updating much simpler. With the database approach, changing the address of a student would mean making *one* single change. With the nondatabase approach, in which each student happened to be stored in three different places, the same change of address would mean that *three* changes had to be made in the computer.

   Although eliminating redundancy is the ideal, it is not always possible. Sometimes, for reasons having to do with performance, we might choose to introduce a limited amount of redundancy into a database. But, even in these cases, we would be able to keep the redundancy under tight control, thus obtaining the same advantages. This is why it is technically better to say that we *control* redundancy rather than *eliminate* it.

6. **Consistency.** Suppose an individual student's address were to appear in more than one place. Student 176, for example, might be listed at 926 Meadowbrook at one spot within our data and 2856 Wisner at another. The data in the computer would then be *inconsistent*. Since the potential for this sort of problem is a direct result of redundancy, and since the database approach eliminates (or at least controls) redundancy, there is much less potential for the occurrence of this sort of inconsistency with the database approach.

7. **Integrity.** An **integrity constraint** is a rule that must be followed by data in the database. Here is an example of an integrity constraint: The director number given for any movie must be that of a director who is *already in the database*. A database has **integrity** if the data in it satisfies all integrity constraints that have been established. A good DBMS should provide an opportunity for users to articulate these integrity constraints when they describe the database. The DBMS should then ensure that these constraints are never violated. According to the integrity constraint just articulated, the DBMS should *not allow* us to store data about a given movie if the director number that we enter is not the number of a director whose name is already in the database.

8. **Security.** **Security** is the prevention of access to the database by unauthorized users. A good DBMS has a number of features that help ensure the enforcement of security measures.

9. **Increased productivity.** A DBMS frees the programmers who are writing programs to access a database from having to engage in mundane data manipulation activities thus making the programmers more productive. A good DBMS comes with many features that allow users to gain access to data in the database without having to do any programming at all. This increases the productivity both of programmers, who may not need to write complex programs in order to perform certain tasks, and of nonprogrammers, who may be able to get the results they seek from the data in the database without waiting for a program to be written for them.

10. **Data independence.** The structure of a database often needs to be changed. For example, new user requirements may necessitate the addition of an entity, an attribute, or a relationship, or a change may be required to improve performance. A good DBMS provides **data independence**, which is the property that the structure of a database can change without the programs that access the database having to change. Without data independence, a lot of unnecessary effort can be expended in changing programs to match the new structure of the database. The presence of many programs in the system may make this effort so prohibitive that a decision is made not to change the database. With data independence, the effort of changing all the programs is unnecessary. Thus, when the need arises to change the database, the decision to do so is more likely to be made.

For other perspective on the advantages of database processing, see [9], [11], [12], [13], [14], and [15].

1.4 DISADVANTAGES OF DATABASE PROCESSING

As you would expect, if there are advantages to doing something in a certain way, there are also disadvantages. The area of database processing is no exception. In terms of numbers alone, the advantages outweigh the disadvantages, but the latter are listed in Figure 1.10 and explained below.

Figure 1.10

Disadvantages of database processing

Disadvantages of Database Processing

1. Larger Size
2. Greater Complexity
3. Greater Impact of a Failure
4. Recovery More Difficult

1. **Larger size.** In order to support all the complex functions that it provides to users, a database management system must be a large program that occupies megabytes of disk space as well as a substantial amount of internal memory.

2. **Greater complexity.** The complexity and breadth of the functions furnished by a DBMS make it a complex product. Users of the DBMS must understand the features of the system in order to take full advantage of it, and there is a great deal for them to learn. In design and implementation of a new system that uses a DBMS, many choices have to be made, and it is possible to make incorrect choices, especially with an insufficient understanding of the system. Unfortunately, a few incorrect choices can spell disaster for the whole project. This is especially true for a large mainframe project that serves many users, but it can also apply to microcomputer projects.

3. **Greater impact of a failure.** If each user has a completely separate system, the failure of any single user's system does not necessarily affect any other user. If, on the other hand, several users are sharing the same database, a failure on the part of any one user which damages the database in some way may affect *all of the other users*.

4. **Recovery more difficult.** Because a database is inherently more complex than a simple file, the process of recovering it in the event of a catastrophe is also more complicated than the process of recovering a simple file. This is particularly true if the database is being updated by a large number of users at the same time. It must first be restored to the condition it was in when it was last known to be correct; any updates made by users since that time must be redone. The greater the number of users involved in updating the database, the more complicated this task becomes.

For other perspectives on the disadvantages of database processing, see [9], [11], [12], [13], [14], and [15].

SUMMARY

1. An entity is a person, place, or thing. An attribute is a property of an entity. A relationship is an association between entities.
2. A database is a structure that can house information about many different entities and about the relationships between these entities.
3. A database management system is a software package whose function is to manipulate a database on behalf of users.
4. Database processing offers a number of advantages, including the following:
   a. lower cost
   b. getting more information from the same amount of data
   c. sharing of data
   d. balancing conflicting requirements
   e. controlled or eliminated redundancy
   f. consistency
   g. integrity
   h. security
   i. increased productivity
   j. data independence
5. The disadvantages to database processing include the following:
   a. larger size
   b. greater complexity
   c. greater impact of a failure
   d. recovery more difficult

KEY TERMS

application package	database management	record
application program	system (DBMS)	redundancy
attribute	entity	relational model
column	field	relationship
data independence	file	security
database	integrity	software package
database design	integrity constraint	table

EXERCISES

1. What is a software package? What is another name for it?
2. What is the difference between an application package and an application program?
3. What is a file? A record? A field?
4. Was Henry implementing a database with the programs he wrote in BASIC? Explain your answer.
5. Give two reasons why Pat did not want to use a DBMS by herself to develop a system for managing her office.
6. What is an entity? An attribute?
7. When we speak of relationship, what exactly do we mean?
8. What is a one-to-many relationship?
9. What is a database?
10. What is a DBMS?
11. Why is the cost of a DBMS listed as an advantage in the microcomputer environment and a disadvantage on mainframes?
12. How is it possible to get more information from the same amount of data through using a database approach as opposed to a file approach?
13. What is meant by "sharing of data"?
14. What is DBA? What kinds of things does DBA do in a database environment?
15. What is redundancy? What are the problems associated with redundancy?
16. How does consistency result from controlling or eliminating redundancy?
17. What is meant by integrity as it is used in this chapter?
18. What is meant by security? What does the DBMS have to do with security?
19. What is meant by data independence? Why is it desirable?
20. How can the size of a DBMS be one of its disadvantages?
21. How can the complexity of a DBMS be a disadvantage?
22. Why can a failure in a database environment be more serious than one in a file environment?
23. Why can recovery be more difficult in a database environment?

2

Data Models

OBJECTIVES

1. Introduce Premiere Products, the company that is used as a basis for many of the examples throughout the text.
2. Introduce the concept of a data model.
3. Describe the relational model.
4. Introduce the network and CODASYL models.
5. Introduce the hierarchical model.

2.1 INTRODUCTION

A database management system (DBMS) must furnish a method for storing and manipulating information about a variety of entities, the attributes of the entities, and the relationships between the entities. Several approaches can be taken by a DBMS to do this, and DBMSs are often categorized by the general approach that they take. This chapter discusses the general categories of DBMSs. These categories are usually called data models.

Since a DBMS must both store and manipulate data, both of these facets must be addressed in a data model. Thus, a **data model** has two components, usually called structure and operations. The **structure** refers to the way in which the system constructs, or structures, the data. More properly, it refers to the way in which the users *perceive* that the system structures data. It really doesn't matter what the DBMS does with the data behind the scenes; it just matters how it appears to the user. The **operations** are the facilities given to the users of the DBMS for the purpose of manipulating data within the database.

The vast majority of DBMSs follow one of three models: the relational model, the network model, or the hierarchical model. In this chapter, we will discuss the basic ideas behind these three models as well as the relative strengths and weaknesses of each one. In subsequent chapters, we will focus only on the relational model. There are two reasons for this. First, the main focus of the book is microcomputer DBMSs, which are almost exclusively relational. Second, even on mainframes, relational model systems are becoming more prevalent all the time. At the very least, systems that follow one of the other two models are typically adding relational characteristics to their systems.

Before introducing the three models, we will examine the requirements of a company called Premiere Products, which will be referred to in many examples throughout this chapter and in the rest of the text. After this examination is completed in section 2.2, we will move on to study the models. We will look at the basic concepts of the relational model in section 2.3, of the network model in section 2.4, and of the hierarchical model in section 2.5.

2.2 PREMIERE PRODUCTS

Premiere Products, a distributor of appliances and sporting goods, needs to maintain the following information:

1. For sales reps, it needs to store the sales rep's number, name, address, total commission, and commission rate.
2. For customers, it needs to store the customer's number, name, address, current balance, and credit limit, and the number of the sales rep who represents the customer.
3. For parts, it needs to store the part's number, its description, the units on hand, the item class, the number of the warehouse in which the item is stored, and the unit price.

Premiere Products also must store information on orders. A sample order is shown in Figure 2.1. Note that there are three parts to the order. The heading (top) of the order contains the order number; the date; the customer's number, name, and address; the sales rep number; and the sales rep name. The body of the order contains a number of *order lines*, sometimes called *line items*. Each order line contains a part number, a part description, the number of units of the part that were ordered, and the quoted price for the part. It also contains a total (usually called an *extension*) which is the product of the number ordered and the quoted price. Finally, the footing (bottom) of the order contains the order total. The additional items that Premiere Products must store with respect to orders are as follows.

Figure 2.1

Premiere Products order

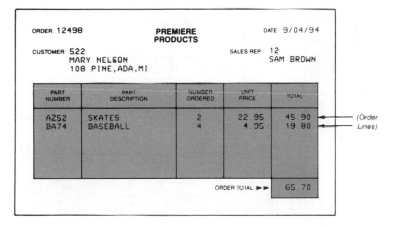

1. For the orders themselves, it needs to store the order number, the date the order was placed, and the number of the customer who placed the order. Note that the customer's name and address and the number of the sales rep who represents the customer are stored with customer information. The name of the sales rep is stored with sales rep information.
2. For each order line, it needs to store the order number, the part number, the number of units ordered, and the quoted price. The part description, you will recall, is stored with information on parts. The product of the number of units ordered and the quoted price is not stored, since it can easily be computed when needed.
3. The overall order total is not stored but will be computed when the order is produced.

Figure 2.2 shows sample data for *PREMIERE PRODUCTS*. We see that there are three sales reps whose numbers are 3, 6, and 12. The name of sales rep 3 is Mary Jones. Her address is 123 Main Street, Grant, MI. Her total commission is $2,150.00, and her commission rate is 5 percent (.05).

Figure 2.2

PREMIERE PRODUCTS sample data

SLSREP

SLSRNUMB	SLSRNAME	SLSRADDR	TOTCOMM	COMMRATE
3	MARY JONES	123 MAIN,GRANT,MI	2150.00	.05
6	WILLIAM SMITH	102 RAYMOND,ADA,MI	4912.50	.07
12	SAM BROWN	419 HARPER,LANSING,MI	2150.00	.05

CUSTOMER

CUSTNUMB	CUSTNAME	ADDRESS	BALANCE	CREDLIM	SLSRNUMB
124	SALLY ADAMS	481 OAK,LANSING,MI	418.75	500	3
256	ANN SAMUELS	215 PETE,GRANT,MI	10.75	800	6
311	DON CHARLES	48 COLLEGE,IRA,MI	200.10	300	12
315	TOM DANIELS	914 CHERRY,KENT,MI	320.75	300	6
405	AL WILLIAMS	519 WATSON,GRANT,MI	201.75	800	12
412	SALLY ADAMS	16 ELM,LANSING,MI	908.75	1000	3
522	MARY NELSON	108 PINE,ADA,MI	49.50	800	12
567	JOE BAKER	808 RIDGE,HARPER,MI	201.20	300	6
587	JUDY ROBERTS	512 PINE,ADA,MI	57.75	500	6
622	DAN MARTIN	419 CHIP,GRANT,MI	575.50	500	3

ORDERS

ORDNUMB	ORDDTE	CUSTNUMB
12489	90294	124
12491	90294	311
12494	90494	315
12495	90494	256
12498	90594	522
12500	90594	124
12504	90594	522

ORDLNE

ORDNUMB	PARTNUMB	NUMBORD	QUOTPRCE
12489	AX12	11	14.95
12491	BT04	1	402.99
12491	BZ66	1	311.95
12494	CB03	4	175.00
12495	CX11	2	57.95
12498	AZ52	2	22.95
12498	BA74	4	4.95
12500	BT04	1	402.99
12504	CZ81	2	108.99

PART

PARTNUMB	PARTDESC	UNONHAND	ITEMCLSS	WREHSENM	UNITPRCE
AX12	IRON	104	HW	3	17.95
AZ52	SKATES	20	SG	2	24.95
BA74	BASEBALL	40	SG	1	4.95
BH22	TOASTER	95	HW	3	34.95
BT04	STOVE	11	AP	2	402.99
BZ66	WASHER	52	AP	3	311.95
CA14	SKILLET	2	HW	3	19.95
CB03	BIKE	44	SG	1	187.50
CX11	MIXER	112	HW	3	57.95
CZ81	WEIGHTS	208	SG	2	108.99

We also see that there are ten customers, numbered 124, 256, 311, 315, 405, 412, 522, 567, 587, and 622. The name of customer 124 is Sally Adams. Her address is 481

Oak Street, Lansing, MI. Her current balance is $418.75, and her credit limit is $500. The number 3 in the column entitled *SLSRNUMB* indicates that Sally is represented by sales rep 3 (Mary Jones).

Skipping down for a moment to the table labeled *PART*, we see that there are ten parts, whose part numbers are AX12, AZ52, BA74, BH22, BT04, BZ66, CA14, CB03, CX11, and CZ81. Part AX12 is an iron, and the company has 104 units of this part on hand. These parts are in item class HW (housewares) and are stored in warehouse 3. The price of an iron is $17.95.

Moving back up to the table labeled *ORDER*, we see that there are seven orders, numbered 12489, 12491, 12494, 12495, 12498, 12500, and 12504. Order 12489 was placed on September 2, 1994, by customer 124 (Sally Adams).

The table labeled *ORDLNE* may seem strange at first glance. Why do we need a separate table for the order lines? Couldn't they somehow be included in the *ORDER* table? The answer is yes, they could. The table *ORDER* could potentially be structured in the manner shown in Figure 2.3. Examining this table, we see that the same orders as those shown in Figure 2.2 are present, with the same dates and the same customer numbers. In addition, each row contains all of the order lines for a given order. Examining the fifth row, for example, we see that order 12498 has two order lines. One of these order lines is for two AZ52's at $22.95 each; the other is for four BA74's at $4.95 each.

Figure 2.3

ORDER and
ORDLNE combined

ORDERS

ORDNUMB	ORDDTE	CUSTNUMB	PARTNUMB	NUMBORD	QUOTPRCE
12489	90294	124	AX12	11	14.95
12491	90294	311	BT04	1	402.99
			BZ66	1	311.95
12494	90494	315	CB03	4	175.00
12495	90494	256	CX11	2	57.95
12498	90594	522	AZ52	2	22.95
			BA74	4	4.95
12500	90594	124	BT04	1	402.99
12504	90594	522	CZ81	2	108.99

Question: How is the same information represented in Figure 2.2?

Answer: Take a look at the table in Figure 2.2 labeled *ORDLNE* and examine the sixth and seventh rows. The sixth row indicates that there is an order line on order 12498 for two AZ52's at $22.95 each. The seventh row indicates that there is an order line on order 12498 for four BA74's at $4.95 each. Thus, the same information that we find in Figure 2.3 is represented here, although in two separate rows rather than one.

It may seem that it would be better not to take two rows to represent the same information that can be represented in one row. There is a problem with the arrangement shown in Figure 2.3, however: the table is more complicated. In Figure 2.2, there is a single entry at each location in the table. In Figure 2.3, some of the individual positions within the table contain multiple entries, and, further, there is a correspondence between these entries (in the row for order 12498, it is crucial to know that the AZ52 corresponds to the 2 in the *NUMBORD* column, not the 4, and the 22.95 in the *QUOTPRCE* column, not the 4.95). There are practical issues to worry about, such as:

1. How much room do we allow for these multiple entries?
2. What if an order has more order lines than we have allowed room for?
3. Given a part, how do we determine which orders contain order lines for that part?

Certainly, none of these problems is unsolvable. They do add a level of complexity, however, that is not present in the arrangement shown in Figure 2.2. In the structure shown in that figure, there are no multiple entries to worry about; it does not matter how many order lines exist for any order; and finding all of the orders that contain order lines for a given part is easy (just look for all order lines with the given part number in the *PARTNUMB* column). In general, this simpler structure is preferable, and that is why order lines have been placed in a separate table.

To test your understanding of the *PREMIERE PRODUCTS* data, answer the following questions, using the data in Figure 2.2.

Q & A

Question: Give the numbers of all the customers represented by MARY JONES.

Answer: 124, 412, and 622. (Look up the number of MARY JONES in the *SLSREP* table and obtain the number 3. Then find all customers in the *CUSTOMER* table that have the number 3 in the *SLSRNUMB* column.)

Question: Give the name of the customer who placed order 12491, then give the name of the sales rep who represents this customer.

Answer: DON CHARLES, SAM BROWN. (Look up the customer number in the *ORDER* table and obtain the number 311. Then find the customer in the *CUSTOMER* table who has customer number 311. Using this customer's sales rep number, which is 12, find the name of the sales rep in the *SLSREP* table.)

Question: List all of the parts that appear on order 12491. For each part, give the description, number ordered, and quoted price.

Answer: *PARTNUMB*: BZ66, *PARTDESC*: WASHER, *NUMBORD*: 1, *QUOTPRCE*: 311.95. Also *PARTNUMB*: BT04, *PARTDESC*: STOVE, *NUMBORD*: 1, *QUOTPRCE*: 402.99. (Look up each *ORDLNE* table row in which the order number is 12491. Each of these rows contains a part number, the number ordered, and the quoted price. The only thing missing is the description of the part. Use the part number to look up the corresponding description in the *PART* table.)

Q & A

Question:	Why is the column *QUOTPRCE* a part of the *ORDLNE* table? Can't we just take the part number and look up the price in the *PART* table?
Answer:	If we do not have the *QUOTPRCE* column in the *ORDLNE* table, the price for a part on an order line must be obtained by looking up the price in the *PART* table. While this may not be bad, it does prevent Premiere Products from charging different prices to different customers for the same part. Since Premiere Products wants the flexibility to quote different prices to different customers, we include the *QUOTPRCE* column in the *ORDLNE* table. If you examine the *ORDLNE* table, you will see cases in which the quoted price matches the actual price in the *PART* table and cases in which it differs.

2.3   THE RELATIONAL MODEL

Structure Within the Relational Model

You have actually already seen a relational database. A relational database is essentially just a collection of tables like the ones we just looked at for Premiere Products in Figure 2.2. A **relational model** database is perceived by the user as being just such a collection. (Notice again the phrase "perceived by the user," indicating as before that what matters is how things appear to the user, not what is taking place behind the scenes.) You might wonder why this model is not called the "table" model, or something along that line if a database is a collection of tables. Formally, these **tables** are called **relations**, and this is where the model gets its name.

How does a DBMS that follows the relational model handle entities, attributes of entities, and relationships between entities? Entities and attributes are fairly simple. Each entity gets a table of its own. Thus, in the *PREMIERE PRODUCTS* database, there is a table for sales reps, a separate table for customers, and so on. The attributes of an entity become the columns in the table. In the table for sales reps, for example, there is a column for the sales rep number, a column for the sales rep name, and so on.

What about relationships? At Premiere Products there is a one-to-many relationship between sales reps and customers (each sales rep is related to the many customers he or she represents, and each customer is represented to the one sales rep who represents the customer). How is this relationship implemented in a relational model database? The answer is through common columns in two or more tables. Consider again Figure 2.2. The column *SLSRNUMB* of the sales rep table and the column *SLSRNUMB* of the customer table are used to implement the relationship between sales reps and customers; that is, given a sales rep, we can use these columns to determine all of the customers he or she represents, and, given a customer, we can use these columns to find the sales rep who represents the customer.

Let us now attempt to be a little more precise in our description of a relation. As we discussed, a relation is essentially just a two-dimensional table. If we consider the tables in Figure 2.2, however, we can see that there are certain restrictions we would probably want to place on relations. Each column should have a unique name, and entries within each column should all "match" this column name; that is, if the column name is *CREDLIM*, all entries in that column should in fact be credit limits. Also, each row should be unique. After all, if two rows are absolutely identical, the second row doesn't give us any information that we don't already have. In addition, for maximum flexibility, the ordering of the columns and the rows should be immaterial. Finally, the table will be simplest if each position is restricted to a single entry, that is, if we do not allow multiple entries (often called **repeating groups**) in an individual location in the table. These ideas lead to the following definitions:

Definition: A **relation** is a two-dimensional table in which

1. the entries in the table are single-valued, that is, each location in the table contains a single entry;
2. each column has a distinct name (technically called the attribute name);
3. all of the values in a column are values of the same attribute (that is, all entries must match the column name);
4. the order of columns is immaterial;
5. each row is distinct; and
6. the order of rows is immaterial.

Definition: A **relational database** is a collection of relations.

Note: Later in the text, we will encounter situations in which a structure satisfies all the properties of a relation *except for property 1;* that is, some of the entries contain repeating groups and thus are not single-valued. Such a structure is called an **unnormalized relation**. This jargon is certainly a little strange, in that an unnormalized relation is thus not a relation at all. It is the term that is used for such a structure, however. The table shown in Figure 2.3 is an example of an unnormalized relation.

Each row of a relation is technically called a **tuple**, and each column is technically called an **attribute**. Thus, we have two different sets of terms: relation, tuple, attribute, and table, row, column. There is even a third set. The table could be viewed as a file (in fact, this is how relational databases are often stored, with each relation in a separate file). In this case, we would call the rows records and the columns fields. We now have *three* different sets of terms! Their correspondence is shown below.

Formal Terms	Alternative One	Alternative Two
relation	table	file
tuple	row	record
attribute	column	field

Of these three sets of choices, the one that is becoming the most popular is alternative one: tables, rows, and columns. One reason for its popularity is that it seems the most natural to the nontechnical user. A second reason is that many (if not most) of the commercial relational DBMSs use this set of terms. All the terms are often used interchangeably. In this text, we will use tables, rows, and columns.

It would be nice to have a concise way of indicating the tables and columns in a relational database without having to draw the tables themselves as we did in Figure 2.2. Perhaps we could draw some empty tables, such as those shown in Figure 2.4. This seems rather cumbersome, however. Fortunately, there is a commonly accepted shorthand representation of the structure of a relational database. We merely write the name of the table and then within parentheses list all of the columns in the table. Thus, this sample database consists of:

```
SLSREP (SLSRNUMB, SLSRNAME, SLSRADDR,
           TOTCOMM, COMMRATE)

CUSTOMER (CUSTNUMB, NAME, ADDRESS, BALANCE,
            CREDLIM, SLSRNUMB)

ORDER (ORDNUMB, DATE, CUSTNUMB)

ORDLNE (ORDNUMB, PARTNUMB, NUMBORD,
           QUOTPRCE)

PART (PARTNUMB, PARTDESC, UNONHAND,
         ITEMCLSS, WREHSENM, UNITPRCE)
```

Figure 2.4

PREMIERE PRODUCTS sample structure

SLSREP

SLSRNUMB	SLSRNAME	SLSRADDR	TOTCOMM	COMMRATE

CUSTOMER

CUSTNUMB	CUSTNAME	ADDRESS	BALANCE	CREDLIM	SLSRNUMB

ORDERS

ORDNUMB	ORDDTE	CUSTNUMB

ORDLNE

ORDNUMB	PARTNUMB	NUMBORD	QUOTPRCE

PART

PARTNUMB	PARTDESC	UNONHAND	ITEMCLSS	WREHSENM	UNITPRCE

Notice that there is some duplication of names. The column *SLSRNUMB* appears in *both* the *SLSREP* table *and* the *CUSTOMER* table. Suppose a situation existed wherein the two might be confused. If we merely wrote *SLSRNUMB*, how would the computer know which *SLSRNUMB* we meant? How would a person looking at what we had written know which one we meant, for that matter? We need a mechanism for indicating the one to which we are referring. One common approach to this problem is to write both the table name and the column name, separated by a period. Thus, the *SLSRNUMB* in the *CUSTOMER* table would be written *CUSTOMER.SLSRNUMB*, whereas the *SLSRNUMB* in the *SLSREP* table would be written *SLSREP.SLSRNUMB*. Technically, when we do this we say that we **qualify** the names. It is *always* acceptable to qualify data names, even if there is no possibility of confusion. If confusion may arise, however, it is *essential* to do so.

There is one other important topic to discuss before we leave the structure part of the relational model. The **primary key** of a table (relation) is the column or collection of columns that uniquely identifies a given row. In the *SLSREP* table, for example, the sales rep's number uniquely identifies a given row. (Sales rep 6 occurs in only one row of the table, for instance.) Thus, *SLSRNUMB* is the primary key. Primary keys are typically indicated in the shorthand representation by underlining the column or collection of columns that comprises the primary key. Thus, the complete shorthand representation for the *PREMIERE PRODUCTS* database would be:

```
SLSREP (SLSRNUMB, SLSRNAME, SLSRADDR,
            TOTCOMM, COMMRATE)

CUSTOMER (CUSTNUMB, NAME, ADDRESS, BALANCE,
            CREDLIM, SLSRNUMB)

ORDER (ORDNUMB, DATE, CUSTNUMB)

ORDLNE (ORDNUMB, PARTNUMB, NUMBORD,
            QUOTPRCE)

PART (PARTNUMB, PARTDESC, UNONHAND,
            ITEMCLSS, WREHSENM, UNITPRCE)
```

Q & A

Question: Why does the primary key to the *ORDLNE* table consist of two columns, not just one?

Answer: No single column uniquely identifies a given row. It requires two: *ORDNUMB* and *PARTNUMB*.

Operations Within the Relational Model

There are many approaches to manipulating a relational database. One of the most prevalent is a language called **SQL (Structured Query Language)**, which was developed by IBM. The basic form of an SQL command is simply

```
SELECT ...
    FROM ...
    WHERE ...
```

We list the columns that we wish to see printed after the word SELECT. After the word FROM, we list all tables that contain these columns. Finally, we list any restrictions to be applied after the word WHERE. For example, if we wanted to print the name of the customer whose number was 124, we would type:

```
SELECT NAME
    FROM CUSTOMER
    WHERE CUSTNUMB = 124
```

and the computer would respond with:

```
NAME

SALLY ADAMS
```

Suppose we wish to see the number and name of each customer together with the number and name of the sales rep who represents the customer. This task involves the use of the two tables, *CUSTOMER* and *SLSREP*, together with the relationship between them. In the SELECT statement, we list all of the columns that we wish included on the report, *CUSTNUMB*, *NAME*, *SLSRNUMB*, and *SLSRNAME*. We have a slight problem, however. There is a *SLSRNUMB* column in both tables, so we must indicate which one we want by qualifying *SLSRNUMB*. (Actually, in this case, since we are looking for combinations of customers and sales reps where the sales rep numbers match, the value of *CUSTOMER.SLSRNUMB* and the value of *SLSREP.SLSRNUMB* will be the same. Thus, it would seem that we shouldn't have to bother with qualifying *SLSRNUMB*. The computer doesn't know that the two will be the same, however, and thus will insist that *SLSRNUMB* be qualified. We would get an error message if we did not do so.) The SELECT clause is thus:

```
SELECT CUSTNUMB, NAME, SLSREP.SLSRNUMB, SLSRNAME
```

After the word FROM, we list both of the tables involved, *CUSTOMER* and *SLSREP*. The FROM clause is:

```
FROM CUSTOMER, SLSREP
```

Finally, we list any restrictions in the WHERE clause. If we did not list any restrictions in the query, we would get *all possible combinations of customers and sales reps*. Therefore, we need a condition that will restrict the output to only those combinations which *match*, that is, the combinations in which *CUSTOMER.SLSRNUMB = SLSREP.SLSRNUMB*. The WHERE clause to accomplish this is:

```
WHERE CUSTOMER.SLSRNUMB = SLSREP.SLSRNUMB
```

The complete query is:

```
SELECT CUSTNUMB, NAME, SLSREP.SLSRNUMB, SLSRNAME
    FROM CUSTOMER, SLSREP
    WHERE CUSTOMER.SLSRNUMB = SLSREP.SLSRNUMB
```

to which the computer would respond:

CUSTNUMB	NAME	SLSRNUMB	SLSRNAME
124	SALLY ADAMS	3	MARY JONES
256	ANN SAMUELS	6	WILLIAM SMITH
311	DON CHARLES	12	SAM BROWN
315	TOM DANIELS	6	WILLIAM SMITH
405	AL WILLIAMS	12	SAM BROWN
412	SALLY ADAMS	3	MARY JONES
522	MARY NELSON	12	SAM BROWN
567	JOE BAKER	6	WILLIAM SMITH
587	JUDY ROBERTS	6	WILLIAM SMITH
622	DAN MARTIN	3	MARY JONES

Advantages and Disadvantages

One big advantage of relational model systems is that they are generally easier to use than systems that follow the other models. A variety of methods are available for manipulating relational databases. These methods are simpler than the ones available for manipulating network and hierarchical model systems. Users of relational systems do not need the in-depth knowledge of the underlying database structure which is required in these other systems.

Another advantage concerns data independence. Data independence, which is the ability to make changes in the database structure without having to make changes in programs that access the database, is one of the advantages of all types of DBMSs. Relational model systems offer a degree of data independence that is much higher than that of network or hierarchical systems. A far wider range of types of changes can be made without affecting programs that access the database. In addition, even those types of changes which can be made using network or hierarchical systems can usually be made much more easily in relational systems.

At the present time, relational systems have two problems, both of which should be corrected in the not-too-distant future. The first problem concerns efficiency. In spite of claims to the contrary, current relational model systems are not as efficient as some of the top network and hierarchical systems. Basically, this means that relational model systems may not be appropriate for developing some large-scale application systems. When there is a great deal of activity involving a very large database, a relational model system may not be able to provide the required performance.

You will probably be surprised at the second problem. It concerns **integrity**, which is the property of ensuring that the data in the database satisfies certain restrictions. These restrictions, which are often called **integrity constraints**, could include such things as the following:

- No two sales reps can have the same number.
- A credit limit must be $300, $500, $800, or $1000.
- The sales rep number on any row in the *CUSTOMER* table must be the number of a sales rep who actually exists, that is, who appears on some row in the *SLSREP* table.

The problem is that many current relational model systems *do not contain any facilities to enforce such integrity constraints*.

For additional perspectives on the basic relational model, see [8], [9], [11], [12], [13], [14], and [15].

2.4 THE NETWORK MODEL

Structure Within the Network Model

A **network model** database is perceived by the user as a collection of record types (which represent the entities), fields within these record types (which represent the attributes), and *explicit* relationships between these record types. Such a structure is called a **network**, and it is from this term that the model takes its name. The fact that the relationships are explicit distinguishes the network model from the relational model, in which relationships are *implicit* (derived from matching columns in the tables).

The CODASYL approach, also called the **CODASYL model**, falls within the general network model. While some systems that are network model systems do not follow the CODASYL model, they are the exception. To many people, the CODASYL model and the network model are synonymous, and that is how we will treat them here. In fact, in the rest of the discussion of the network model, we will use CODASYL terminology.

Consider the database in Figure 2.5, for example. The rectangles represent the record types in the database. There is one for each of the entities: sales reps, customers, orders, parts, and order lines.

Figure 2.5

PREMIERE PRODUCTS network database structure

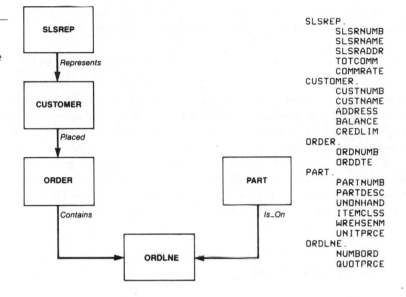

```
SLSREP.
    SLSRNUMB
    SLSRNAME
    SLSRADDR
    TOTCOMM
    COMMRATE
CUSTOMER.
    CUSTNUMB
    CUSTNAME
    ADDRESS
    BALANCE
    CREDLIM
ORDER.
    ORDNUMB
    ORDDTE
PART.
    PARTNUMB
    PARTDESC
    UNONHAND
    ITEMCLSS
    WREHSENM
    UNITPRCE
ORDLNE.
    NUMBORD
    QUOTPRCE
```

The arrows represent the relationships. In particular, they represent the familiar one-to-many relationships. The arrow goes from the "one" part of the relationship, called the **owner**, to the "many" part, called the **member**. Since one sales rep is related to many customers but each customer is related to exactly one sales rep, we have a one-to-many relationship, and thus an arrow, from *SLSREP* to *CUSTOMER*. In this relationship, *SLSREP* is the owner and *CUSTOMER* is the member. There is a term for the relationship in CODASYL systems; it is called a **set**. (Note that this use of the term differs from the one usually associated with it in mathematics.)

Sets are given names. The set from *SLSREP* to *CUSTOMER* is called *REPRESENTS*, indicative of the fact that a sales rep "represents" a customer. The set from *CUSTOMER* to *ORDER* is called *PLACED*, to indicate that a customer "placed" a number of orders. Similarly, an order "contains" order lines and a part "is on" a number of order lines, so the set from *ORDER* to *ORDLNE* is called *CONTAINS* and the set from *PART* to *ORDLNE* is called *IS_ON*.

Operations Within the Network Model

As we will see in the following examples, we manipulate a network model database essentially by *following the arrows*. Arrows may be followed in either direction. Suppose we locate a given sales rep, for example. If we follow the arrow from *SLSREP* to *CUSTOMER*, we obtain the many customers represented by this sales rep. If we start with a customer, however, and follow the same arrow only in the reverse direction, we obtain the single sales rep who represents the given customer, that is, the single owner.

The sample data shown in Figure 2.6 illustrates these processes. This figure shows three sales reps and ten customers. The relationship between them is indicated by the lines in the figure. Thus, sales rep 3 represents customers 124, 412, and 622; sales rep 6 represents customers 256, 315, 567, and 587; sales rep 12 represents customers 311, 405, and 522. (While we will not go into the manner in which these relationships are actually implemented, this is a reasonable way to picture it.) The following three examples demonstrate the process of manipulating a network model database.

Example 1: Given the data shown in Figure 2.6, print a list of all of the customers of sales rep 12.

We first ask the DBMS to FIND sales rep 12. Assuming there is such a sales rep in the database, we then repeatedly ask to FIND the NEXT customer within the collection of customers represented by this sales rep, until reaching the end of the list of such customers. (The exact command would be *FIND NEXT WITHIN REPRESENTS*.) The first time we made the request, we would obtain customer 311; the second time, we would obtain customer 405; the next time, we would obtain customer 522. When we tried to find another customer related to this sales rep, the DBMS would indicate that there were no more. At this point, our task would be complete.

Figure 2.6

Sales reps,
customers, and the
relationship
between them

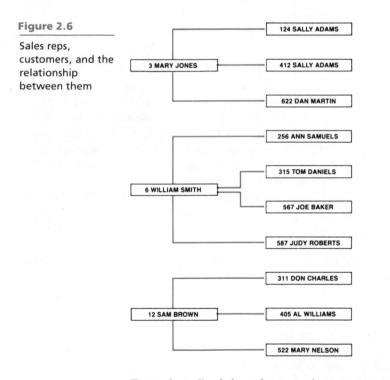

Example 2: Find the sales rep who represents customer 315.

We first ask the DBMS to FIND customer 315. Having accomplished this, we then ask the DBMS to FIND the OWNER of customer 315 in the relationship between sales reps and customers. The owner is a *single* sales rep, in this example, sales rep 6. (The exact command is *FIND OWNER WITHIN REPRESENTS*.)

Note that we find the sales rep who represents a customer by following the arrow, not by looking at a sales rep number in a customer record, as we would if we were using the relational model. Therefore, in a network structure, there is *no need for such a field*.

Some processing involves following more than a single arrow. Consider Figure 2.7, for example. This figure contains the same data and relationships that were shown in Figure 2.6. Additionally, it contains orders, together with the relationships between customers and orders. Thus we see that customer 124 placed orders 12489 and 12500; customer 256 placed order 12495; and so on.

Figure 2.7

Sales reps, customers, orders, and the relationship between them

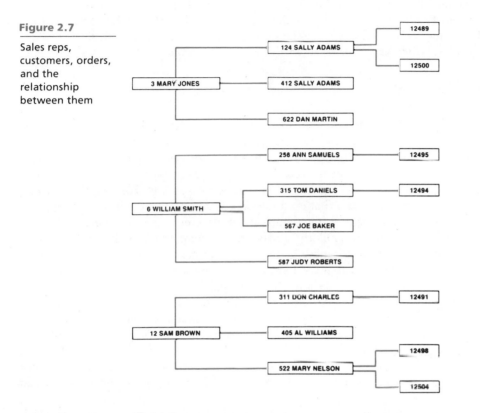

Example 3: List all the orders placed by customers of sales rep 12.

As in Example 1, we first ask the DBMS to FIND sales rep 12. We also repeatedly ask to FIND the NEXT customer within the collection of customers represented by this sales rep (*FIND NEXT WITHIN REPRESENTS*) until reaching the end of the list of such customers. The difference here is that once we have found a customer, we repeatedly ask the DBMS to FIND the NEXT order owned by the customer (*FIND NEXT WITHIN PLACED*). We do this until reaching the end of the list of orders placed by the indicated customer. Only at this point do we proceed on to the next customer owned by the sales rep. In our example, after finding customer 311, we would obtain the single order 12491. Since this is the only order for this customer, we would then be able to proceed to customer 405. Since this customer has no orders, we would then be able to proceed on to customer 522. At this point, we would find the first order for this customer, order 12498. After displaying details about this order, we would move on to the next order placed by this customer, order 12504. Since this is the last order for this customer, once we had displayed details about this order, we would be ready to move on to the next customer. There are, however, no more customers, and thus the whole process would then be complete.

Q & A

Question: How would you handle the following query: List the number and name of the sales rep who represents customer 124. In addition, list the numbers of all orders placed by this customer.

Answer: Ask the DBMS to FIND customer 124. Assuming this step is successfully completed, ask the DBMS to FIND the OWNER of this customer in the relationship between sales reps and customers (*FIND OWNER WITHIN REPRESENTS*). Finally, ask the DBMS to repeatedly FIND the NEXT order owned by this customer (*FIND NEXT WITHIN PLACED*).

Advantages and Disadvantages

The biggest advantage of network model systems is efficiency. Network model DBMSs are known for their ability to handle huge databases with large amounts of activity. Network model systems have also existed for some time now, and a great deal of software has been developed which uses them. Network systems also provide some of the types of integrity support which are lacking in current relational systems. For example, in network systems, it is easy to ensure that each sales rep in the database will have a different number; if sales rep 3 is already in the database, an attempt to add a second sales rep 3 will be reflected by the DBMS. Similarly, a network system guarantees that the data for a sales rep will be placed in the database before any customer he or she represents. Thus, we know that if Dan Martin is assigned to Mary Jones, who is sales rep 3, he will not be placed in the database unless she has already been placed there.

Network systems, on the other hand, are more difficult to use than relational systems. Getting the desired information out of a network database involves following the proper sequence of arrows in the correct order. We need to be very familiar with the underlying database structure in order to determine how best to work our way through the database in order to satisfy a given requirement in an efficient manner. If we do not do this correctly, we will not obtain the additional efficiency the network model can provide.

It is also more difficult to make changes in the database structure in a network system than in a relational system. Many types of changes that could be made in relational systems without affecting application programs cannot be made in any easy way in network systems, and, if they are made, changes are required in all of the programs that access the affected part of the database.

For additional perspectives on the general network model and the CODASYL model, see [9], [10], [11], [13], [14], and [15].

2.5   THE HIERARCHICAL MODEL

Structure and Operations Within the Hierarchical Model

Where the structure used in the network model is the network, the structure used in the **hierarchical model** is the **hierarchy**, or **tree**. A tree can be thought of as a network with an added restriction: no box can have more than one arrow entering it. (It doesn't matter how many arrows leave a box.) A tree is thus a more restrictive structure than a network.

Some of the terminology used for the two models differs as well. Rather than the "one" and "many" ends of a relationship being called *owners* and *members*, as they are in the network model, they are called **parents** and **children** in the hierarchical model.

There is no special name for a relationship in the hierarchical model; that is, there is no term that is analogous to the term *set* in the network model. A different collection of commands is used to manipulate hierarchical model databases, although the basic ideas are similar. There are differences in the underlying strategies used by hierarchical and network DBMSs to store and manipulate the database. Perhaps the biggest difference, however, arises when the database to be implemented is not a tree.

Consider the network shown in Figure 2.8, for example. In this network, a box called *PAYMENT* has been added to the diagram shown in Figure 2.5. This new box will be used to keep track of the payments made by each customer. There is a one-to-many relationship between *CUSTOMER* and *PAYMENT*, which is represented by an arrow from *CUSTOMER* to *PAYMENT*. Thus, in this new diagram, there are two arrows *leaving* the *CUSTOMER* box. There are also two arrows *entering* the *ORDLNE* box. The two arrows leaving the *CUSTOMER* box do not prevent this structure from being a tree. The fact that there is a box, namely *ORDLNE*, with two arrows entering it prevents it from being a tree. The structure shown in Figure 2.8, which is thus not a tree, *cannot* be implemented directly by means of a hierarchical system.

Figure 2.8

PREMIERE PRODUCTS network database structure with additional record type (*PAYMENT*)

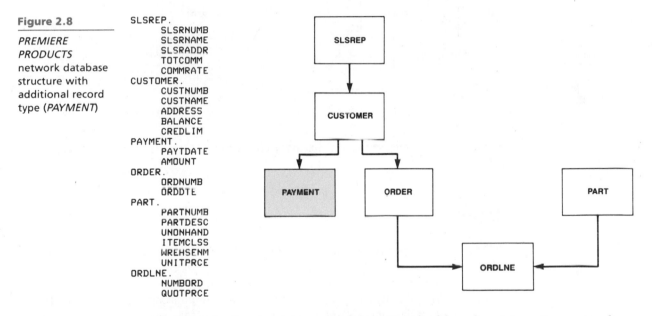

Fortunately, the structure can be implemented, although not in quite as appealing a fashion as is possible in a network system. Although many of the issues involved in such an implementation are rather tricky and are beyond the scope of our discussion, the basic idea is relatively simple. The idea is to create two separate trees (each called a **physical database**), and to create a relationship in which the parent is in one of the physical databases and the child is in the other. Such a relationship is called a **logical child relationship**.

Figure 2.9 shows how this technique could be applied to the structure of Figure 2.8. One physical database contains sales reps, customers, payments, orders, and order lines, as well as all of the relationships between them. The other contains only parts. In addition, the relationship between parts and order lines is implemented as a logical child relationship, in which the *PART* record is the parent and the *ORDLNE* record is the child.

Advantages and Disadvantages

Relative to relational model systems, hierarchical systems offer the same advantages and suffer from the same disadvantages that network systems do. In many ways, hierarchical systems and network systems are comparable. If the underlying database design is really a tree, then a hierarchical model system may outperform a network system. On the other hand, if the underlying structure is not a tree, the reverse may very well be true. In general, though, the systems are quite comparable, so the advantages and disadvantages discussed for the network model apply equally well here.

Figure 2.9

PREMIERE PRODUCTS database split into two separate physical databases with a logical child relationship between them

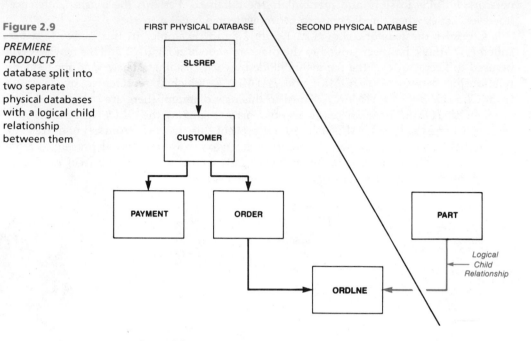

For additional perspectives on the hierarchical model, see [9], [10], [11], [13], [14], and [15].

SUMMARY

1. DBMSs are categorized by the data model that they follow. A data model has two components:
   a. structure (the way the system structures the data)
   b. operations (the facilities provided to users for manipulating data in the database)
2. The three main data models are as follows:
   a. the relational model
   b. the network model
   c. the hierarchical model
3. Premiere Products is an organization whose requirements include the following entities:
   a. sales reps   d. parts
   b. customers   e. order lines
   c. orders
4. A relation is a two-dimensional table in which
   a. the entries are single-valued;
   b. each column has a distinct name;
   c. all of the values in a column are values of the same attribute (the one identified by the column name);
   d. the order of columns is immaterial;
   e. each row is distinct; and
   f. the order of rows is immaterial.

5. A relational database is a collection of relations.
6. An unnormalized relation is a structure in which entries need not be single-valued but which satisfies all of the other properties of a relation.
7. The terms relation, tuple, and attribute correspond to the terms table, row, and column, respectively.
8. Relation, tuple, and attribute also correspond to file, record, and field, respectively.
9. A column name is qualified by preceding it with the table name and a period, for example, *SLSREP.SLSRNUMB*.
10. The primary key is the column or columns that uniquely identify a given row within the table.
11. SQL is a language used to manipulate relational databases. The basic form of an SQL command is SELECT-FROM-WHERE.
12. The advantages of relational model systems are
    a. ease of use and
    b. a high degree of data independence.
13. The disadvantages of relational model systems are that
    a. they are less efficient than systems that follow the other models, and
    b. they are weak in integrity support.
14. A network model database is a collection of record types and explicit one-to-many relationships between these record types.
15. The CODASYL model is a subset of the network model, in which
    a. relationships are called sets;
    b. the "one" part of the set is called the owner; and
    c. the "many" part is called the member.
16. The advantages of network model systems are that
    a. they tend to be more efficient than relational systems, and
    b. they provide better integrity support than relational systems.
17. The disadvantages of network model systems are that
    a. they tend to be more difficult to use than relational systems, and
    b. they provide less data independence than relational systems.
18. The structure in the hierarchical model is the hierarchy.
19. A hierarchy (also called a tree structure) is a network structure with an added restriction: no box can have more than one arrow entering it.
21. If a network that is *not* a tree is to be implemented using a hierarchy system, the network is split into trees with logical child relationships between them.
22. Relative to relational model systems, hierarchical systems have the same advantages and disadvantages that the network systems do.
23. If the structure to be implemented is actually a tree, a hierarchical system may outperform a network system. If it is not a tree, the reverse may very well be the case.

KEY TERMS

children	network model	relational model
CODASYL model	owner	repeating group
data model	parent	row
hierarchical model	physical database	set
hierarchy	primary key	table
logical child	qualification	tree
member	relation	tuple
network	relational database	unnormalized relation

EXERCISES

1. What is a data model?
2. What are the components of a data model?
3. What are the three main data models?
4. Using the data for *PREMIERE PRODUCTS* as shown in Figure 2.2, give an answer for each of the following problems.
   a. Give the names of all customers represented by William Smith.
   b. How many customers have a balance that is over their credit limit?
   c. Which sales reps represent customers whose balance is over their credit limit?
   d. Which customers placed orders on 9/05/94?
   e. Which sales reps represent any customers who placed orders on 9/05/94?
   f. Which customers currently have a STOVE on order?
   g. List the number and description of all parts that are currently on order by any customer represented by William Smith.
5. Why are order lines in the *PREMIERE PRODUCTS* database in a separate table rather than being part of the *ORDERS* table?
6. What is a relation?
7. What is a relational database?
8. What is an unnormalized relation? Is it a relation according to the definition of the word *relation*?
9. What is a tuple? What is a more common name for it?
10. How is the term *attribute* used in the relational model? What is a more common name for it?
11. Describe the shorthand representation of the structure of a relational database. Illustrate this technique by representing Henry's database as shown in Figures 1.1, 1.2, and 1.3.
12. What does it mean to qualify the name of a column? How is this done?
13. What is a primary key? What is the primary key for each of the tables in Henry's database? (See Exercise 11.)
14. What is SQL? What is it used for?
15. What are the advantages of the relational model? What are the disadvantages?
16. What is a network?
17. What is the relationship between the network model and the CODASYL model?
18. What is a set? An owner? A member?
19. Why might there be columns in a relational model table for which there are no corresponding fields in a network model record?
20. How would you handle the following query in a CODASYL model database for *PREMIERE PRODUCTS*: Print a list of all orders placed by customers of sales rep 6.
21. What are the advantages of the network model compared to the relational model? What are the disadvantages?
22. What is another name for hierarchy? What is the difference between a hierarchy and a network?
23. Describe briefly how a network that is not a hierarchy can be implemented by means of a hierarchical model DBMS.
24. What are the advantages of the hierarchical model as compared to the other two models? What are the disadvantages?

The Relational Model: Data Definition and Manipulation

OBJECTIVES

1. Present the language SQL and illustrate data definition and manipulation in SQL.
2. Describe the relational algebra and demonstrate the select, project, and join operations.

3. Present QBE (Query-by-Example) and illustrate its use in retrieving data.
4. Demonstrate the concept of a natural language through the use of a hypothetical natural language called ''NL''.

3.1 INTRODUCTION

In the last chapter, we discussed the basic structure of the relational model. We saw that data is stored in the form of tables. Each entity in the database has a table of its own. The attributes (properties) of the entity are the columns in the table. Relationships between entities are affected through common columns.

In this chapter, we examine a variety of ways of manipulating data in a relational database. In section 3.2, we discuss the language SQL (Structured Query Language). Originally developed by IBM under the name SEQUEL, SQL is perhaps the most important language designed to manipulate relational databases. SQL is not only the language used in DB2, IBM's mainframe relational DBMS; it is also used in many other DBMSs as well, both on mainframes and on microcomputers. A myriad of systems use it already, and there are constant rumors about other systems furnishing this support in the not-too-distant future.

In section 3.3, we will study a very visual approach to the process, called QBE (Query-by-Example). In section 3.4, we will investigate the relational algebra, one of the first methods proposed for retrieving data from relational databases. Finally, in section 3.5, we discuss the use of natural languages to retrieve data from relational databases. The examples used throughout this chapter all refer to the *PREMIERE PRODUCTS* database (see Figure 3.1 on the next page).

Figure 3.1

*PREMIERE
PRODUCTS*
sample data

SLSREP

SLSRNUMB	SLSRNAME	SLSRADDR	TOTCOMM	COMMRATE
3	MARY JONES	123 MAIN,GRANT,MI	2150.00	.05
6	WILLIAM SMITH	102 RAYMOND,ADA,MI	4912.50	.07
12	SAM BROWN	419 HARPER,LANSING,MI	2150.00	.05

CUSTOMER

CUSTNUMB	CUSTNAME	ADDRESS	BALANCE	CREDLIM	SLSRNUMB
124	SALLY ADAMS	481 OAK,LANSING,MI	418.75	500	3
256	ANN SAMUELS	215 PETE,GRANT,MI	10.75	800	6
311	DON CHARLES	48 COLLEGE,IRA,MI	200.10	300	12
315	TOM DANIELS	914 CHERRY,KENT,MI	320.75	300	6
405	AL WILLIAMS	519 WATSON,GRANT,MI	201.75	800	12
412	SALLY ADAMS	16 ELM,LANSING,MI	908.75	1000	3
522	MARY NELSON	108 PINE,ADA,MI	49.50	800	12
567	JOE BAKER	808 RIDGE,HARPER,MI	201.20	300	6
587	JUDY ROBERTS	512 PINE,ADA,MI	57.75	500	6
622	DAN MARTIN	419 CHIP,GRANT,MI	575.50	500	3

ORDERS

ORDNUMB	ORDDTE	CUSTNUMB
12489	90294	124
12491	90294	311
12494	90494	315
12495	90494	256
12498	90594	522
12500	90594	124
12504	90594	522

ORDLNE

ORDNUMB	PARTNUMB	NUMBORD	QUOTPRCE
12489	AX12	11	14.95
12491	BT04	1	402.99
12491	BZ66	1	311.95
12494	CB03	4	175.00
12495	CX11	2	57.95
12498	AZ52	2	22.95
12498	BA74	4	4.95
12500	BT04	1	402.99
12504	CZ81	2	108.99

PART

PARTNUMB	PARTDESC	UNONHAND	ITEMCLSS	WREHSENM	UNITPRCE
AX12	IRON	104	HW	3	17.95
AZ52	SKATES	20	SG	2	24.95
BA74	BASEBALL	40	SG	1	4.95
BH22	TOASTER	95	HW	3	34.95
BT04	STOVE	11	AP	2	402.99
BZ66	WASHER	52	AP	3	311.95
CA14	SKILLET	2	HW	3	19.95
CB03	BIKE	44	SG	1	187.50
CX11	MIXER	112	HW	3	57.95
CZ81	WEIGHTS	208	SG	2	108.99

3.2   SQL

In this section, we examine the SQL language through a number of examples. Like most modern languages, SQL is basically free format. This means that there are no special rules for spacing when one is typing an SQL command. Commas, however, are essential. There is one little problem that is not unique to SQL. The word *ORDER* has special meaning in SQL and cannot be used as the name of a table. In any system with such a restriction, we must pick another name for the *ORDER* table. In our examples, we will use the name *ORDERS*.

In the examples that follow, we will investigate the manner in which tables may be described, data may be retrieved, new data may be added, data may be changed, and data may be deleted.

Data Definition

Example 1: Creation of a database.

Statement: Describe the layout of the sales rep table to the DBMS.

```
CREATE TABLE SLSREP
     (SLSRNUMB          DECIMAL(2),
      SLSRNAME          CHAR(15),
      SLSRADDR          CHAR(25),
      TOTCOMM           DECIMAL(7,2),
      COMMRATE          DECIMAL(2,2))
```

This illustrates the manner in which, using SQL, we describe a new table to the DBMS. We name the table, the columns, and the physical characteristics of the columns. In this example, we are describing a table that will be called *SLSREP*. It contains five columns: *SLSRNUMB, SLSRNAME, SLSRADDR, TOTCOMM,* and *COMMRATE.* The word *DECIMAL* indicates that *SLSRNUMB* is a numeric field; that is, it can contain only numbers. The number 2 in parentheses indicates that *SLSRNUMB* is two digits in length. The *CHAR* indicates that *SLSRNAME* is a character field (also called an *alphanumeric* field); that is, it can contain any type of character. The number 15 indicates that *SLSRNAME* is fifteen characters in length. Similarly, *SLSRADDR* is a character field that is twenty-five characters in length. *TOTCOMM* is numeric and is seven digits long. The 2 that follows the comma indicates that the last two of the seven digits represent two decimal places. Similarly, *COMMRATE* is two digits long, and both of the two digits are decimal places. We can visualize this statement as setting up for us a blank table with appropriate column headings (see Figure 3.2).

Figure 3.2

Blank *SLSREP* table

SLSREP

SLSRNUMB	SLSRNAME	SLSRADDR	TOTCOMM	COMMRATE

We now consider the data manipulation features of the language, beginning with those aspects of the language which are devoted to retrieving information from the database.

Simple Retrieval

The basic form of an SQL expression is quite simple. It is merely SELECT-FROM-WHERE. After the SELECT, we list those columns which we wish to have displayed. After the FROM, we list the table or tables that are involved in the query. Finally, after the WHERE, we list any conditions that apply to the data we wish to retrieve.

Example 2: Retrieval of certain columns and all rows.

Statement: List the number, name, and balance of all customers.

Since we want all customers listed, there is no need for the WHERE clause (we have no restrictions). The query is thus:

```
SELECT CUSTNUMB, CUSTNAME, BALANCE
     FROM CUSTOMER
```

The computer will respond with:

CUSTNUMB	CUSTNAME	BALANCE
124	SALLY ADAMS	418.75
256	ANN SAMUELS	10.75
311	DON CHARLES	200.10
315	TOM DANIELS	320.75
405	AL WILLIAMS	201.75
412	SALLY ADAMS	908.75
522	MARY NELSON	49.50
567	JOE BAKER	201.20
587	JUDY ROBERTS	57.75
622	DAN MARTIN	575.50

Example 3: Retrieval of all columns and all rows.

Statement: List the complete *ORDERS* table.

We could certainly use the same structure that is shown in Example 2. However, a shortcut is available. Instead of listing all of the column names after the SELECT, we can use the symbol *. This indicates that we want all columns listed (in the order in which they have been described to the system during data definition). If we want all columns listed but in a different order, we would have to type the names of the columns in the order in which we wanted them to appear. In this case, assuming that the normal order is appropriate, the query would be:

```
SELECT *
     FROM ORDERS
```

Result:

ORDNUMB	ORDDTE	CUSTNUMB
12489	90294	124
12491	90294	311
12494	90494	315
12495	90494	256
12498	90594	522
12500	90594	124
12504	90594	522

Example 4: Use of the WHERE clause.

Statement: What is the name of customer 124?

We use the WHERE clause to restrict the output of the query to customer 124 as follows:

```
SELECT CUSTNAME
     FROM CUSTOMER
     WHERE CUSTNUMB = 124
```

Result:

CUSTNAME
SALLY ADAMS

The condition in the WHERE clause need not involve equality. We can also use > (greater than), > = (greater than or equal to), < (less than), < =, (less than or equal to), or ~ = (not equal). (In some settings, < > is used for not equal.)

Example 5: Use of a compound condition within the WHERE clause.

Statement: List the descriptions of all parts that are in warehouse 3 and for which there are more than 100 units on hand.

Compound conditions are possible within the WHERE clause using AND, OR, and NOT. In this case, we have:

```
SELECT PARTDESC
      FROM PART
      WHERE WREHSENM = 3
      AND UNONHAND > 100
```

Result:

```
PARTDESC

IRON
MIXER
```

Example 6: Use of computed columns.

Statement: List the number, name, and available credit for all customers who have at least an $800 credit limit.

This statement poses a problem for us. There is no *available credit* column in our database. It is, however, computable from two columns that are present, *CREDLIM* and *BALANCE* (*available credit* = *CREDLIM* − *BALANCE*). Fortunately, SQL permits us to specify computations within the SQL expression. In this case, we would have:

```
SELECT CUSTNUMB, CUSTNAME, CREDLIM - BALANCE
      FROM CUSTOMER
      WHERE CREDLIM >= 800
```

Result:

```
CUSTNUMB CUSTNAME      3

     256 ANN SAMUELS 789.25
     405 AL WILLIAMS 598.25
     412 SALLY ADAMS  91.25
     522 MARY NELSON 750.50
```

Note that the heading for the new column is simply the number 3. Since this column does not exist in the *CUSTOMER* table, the computer does not know how to label the column and instead uses the number 3 (for the *third* column). There is a facility within SQL to change any of the column headings to whatever we desire. For now, though, we will just accept the headings that SQL will produce automatically. (Some variation exists among different versions of SQL regarding column headings for computed columns. Your version may very well treat them differently. Some versions, for example, will label the column as *CREDLIM − BALANCE*; that is, the computation will be used as the column heading.)

Example 7: Use of a wild card.

Statement: List the number, name, and address of all customers who live in Grant.

The only problem posed by this query results from a single column containing the street address, the city, and the state. If we had a separate column for the city, the query would not be difficult (the WHERE clause would be *WHERE CITY = 'GRANT'*). In this case, however, all we can say is that anyone living in Grant has GRANT somewhere within his or her address, but *we don't know where*. Fortunately, SQL provides a facility that we can use in this situation, as the following illustrates:

```
SELECT CUSTNUMB, CUSTNAME, ADDRESS
     FROM CUSTOMER
     WHERE ADDRESS LIKE '%GRANT%'
```

Result:

CUSTNUMB	CUSTNAME	ADDRESS
256	ANN SAMUELS	215 PETE,GRANT,MI
405	AL WILLIAMS	519 WATSON,GRANT,MI
622	DAN MARTIN	419 CHIP,GRANT,MI

The symbol % is used as what is termed a *wild card*. Using this feature, we are asking for all customers whose address is "like" some collection of characters, followed by GRANT, followed by some other characters. Note that this query would also pick up a customer whose address was 123 Grantview Street, Ada, Michigan. We would probably be safer to ask for addresses like '%,GRANT,%'.

Sorting

In any language that is used to access a database, an essential feature is the ability to sort the data that is retrieved in whatever manner is desired. In SQL, this is accomplished by the ORDER BY clause, as the following example demonstrates.

Example 8: Sorting.

Statement: List the number, name, and address of all customers. The report should be ordered by name.

To have the output sorted, indicate the key on which the data is to be sorted in an ORDER BY clause. In this example, the formulation is:

```
SELECT CUSTNUMB, CUSTNAME, ADDRESS
     FROM CUSTOMER
     ORDER BY CUSTNAME
```

Result:

CUSTNUMB	CUSTNAME	ADDRESS
405	AL WILLIAMS	519 WATSON,GRANT,MI
256	ANN SAMUELS	215 PETE,GRANT,MI
622	DAN MARTIN	419 CHIP,GRANT,MI
311	DON CHARLES	48 COLLEGE,IRA,MI
567	JOE BAKER	808 RIDGE,HARPER,MI
587	JUDY ROBERTS	512 PINE,ADA,MI
522	MARY NELSON	108 PINE,ADA,MI
412	SALLY ADAMS	16 ELM,LANSING,MI
124	SALLY ADAMS	481 OAK,LANSING,MI
315	TOM DANIELS	914 CHERRY,KENT,MI

(Note that since the names are stored in a single column and we have stored the names in the form FIRST LAST, the list is alphabetized on the basis of *first* names.)

Example 9: Sorting with multiple keys, descending order.

Statement: List the customer number, name, and credit limit of all customers, ordered by decreasing credit limit and by customer number within credit limit.

This example calls for sorting on multiple keys (customer number within credit limit) as well as the use of descending order for one of the keys. This is accomplished as follows:

```
SELECT CUSTNUMB, CUSTNAME, CREDLIM
     FROM CUSTOMER
     ORDER BY CREDLIM DESC, CUSTNUMB
```

Result:

CUSTNUMB	CUSTNAME	CREDLIM
412	SALLY ADAMS	1000
256	ANN SAMUELS	800
405	AL WILLIAMS	800
522	MARY NELSON	800
124	SALLY ADAMS	500
587	JUDY ROBERTS	500
622	DAN MARTIN	500
311	DON CHARLES	300
315	TOM DANIELS	300
567	JOE BAKER	300

Built-in Functions

Many languages include built-in functions to allow the calculation of the number of entries, the sum or average of all of the entries in a given column, and the largest or smallest of the entries in a given column. In SQL, these functions are called COUNT, SUM, AVG, MAX, and MIN, respectively. The manner in which they are used is demonstrated in the next two examples.

Example 10: Use of the built-in function COUNT.

Statement: How many parts are in item class HW?

In this query, we are interested in the number of rows in the table for which the item class is HW. We could count the number of part numbers in this table or the number of descriptions or the number of entries in any other column; it doesn't make any difference. (Rather than requiring us to arbitrarily pick one of these, some versions of SQL allow us to use the symbol *.) The query could thus be formulated as follows:

```
SELECT COUNT(PARTNUMB)
     FROM PART
     WHERE ITEMCLSS = 'HW'
```

Result:

$$\frac{1}{4}$$

Example 11: Use of SUM.

Statement: Count the number of customers and find the total of their balances.

The only real difference between COUNT and SUM (other than the obvious one that they are computing different statistics) is that in the case of SUM, we *must* specify the column for which we want a total. (It doesn't make sense to use the *, even if this is permitted for COUNT.) This query is thus:

```
SELECT COUNT(CUSTNUMB), SUM(BALANCE)
    FROM CUSTOMER
```

Result:

1	2
10	2944.8

Querying Two Tables

Example 12: Joining two tables together.

Statement: For each customer, list the number and name of the customer together with the number and name of the sales rep who represents the customer.

Here we indicate all columns we wish displayed in the SELECT clause. In the FROM clause, we list all of the tables involved in the query. Finally, in the WHERE clause, we give the condition that will restrict the data to be retrieved to only those rows from the two tables that match. We have a problem, however. There is a column in *SLSREP* called *SLSRNUMB*, as well as a column in *CUSTOMER* called *SLSRNUMB*. In this case, if we merely mention *SLSRNUMB*, SQL will not know which one we mean. It is necessary to **qualify** *SLSRNUMB*, that is, to specify which one we are referring to. You will recall that we do this by preceding the name of the column with the name of the table, followed by a period. (The *SLSRNUMB* column in the *SLSREP* table is *SLSREP.SLSRNUMB*, and the *SLSRNUMB* column in the *CUSTOMER* table is *CUSTOMER.SLSRNUMB*.) The query is as follows.

```
SELECT CUSTNUMB, CUSTNAME, SLSREP.SLSRNUMB, SLSRNAME
    FROM CUSTOMER, SLSREP
    WHERE CUSTOMER.SLSRNUMB = SLSREP.SLSRNUMB
```

Result:

CUSTNUMB	CUSTNAME	SLSRNUMB	SLSRNAME
124	SALLY ADAMS	3	MARY JONES
256	ANN SAMUELS	6	WILLIAM SMITH
311	DON CHARLES	12	SAM BROWN
315	TOM DANIELS	6	WILLIAM SMITH
405	AL WILLIAMS	12	SAM BROWN
412	SALLY ADAMS	3	MARY JONES
522	MARY NELSON	12	SAM BROWN
567	JOE BAKER	6	WILLIAM SMITH
587	JUDY ROBERTS	6	WILLIAM SMITH
622	DAN MARTIN	3	MARY JONES

Example 13: Restricting the rows in a JOIN.

Statement: For each customer whose credit limit is $800, list the number and name of the customer together with the number and name of the sales rep who represents the customer.

In the previous example, the condition in the WHERE clause served only to relate a customer to a sales rep. While relating a customer to a sales rep is essential in this example as well, we also want to restrict the output to only those customers whose credit limit is $800. This is accomplished by a compound condition, as follows:

```
SELECT CUSTNUMB, CUSTNAME, SLSREP.SLSRNUMB, SLSRNAME
    FROM CUSTOMER, SLSREP
    WHERE CUSTOMER.SLSRNUMB = SLSREP.SLSRNUMB
    AND CREDLIM = 800
```

Result:

CUSTNUMB	CUSTNAME	SLSRNUMB	SLSRNAME
256	ANN SAMUELS	6	WILLIAM SMITH
405	AL WILLIAMS	12	SAM BROWN
522	MARY NELSON	12	SAM BROWN

Subqueries

Example 14: Nesting queries.

Statement: List the number and name of any sales rep who represents any customer whose balance exceeds his or her credit limit.

The only real difference between this query and the preceding one is that in this one, we are not required to list any information about the customer. Customer information still enters the picture, however, since it is used in determining which sales reps should be listed. For this reason, the *CUSTOMER* table is still listed in the FROM clause. The query is thus:

```
SELECT SLSREP.SLSRNUMB, SLSRNAME
    FROM CUSTOMER, SLSREP
    WHERE CUSTOMER.SLSRNUMB = SLSREP.SLSRNUMB
    AND BALANCE > CREDLIM
```

Result:

SLSRNUMB	SLSRNAME
3	MARY JONES
6	WILLIAM SMITH

There is another way in which we could logically approach the problem. We could do it in two steps. We could first find the numbers of all the sales reps who represent such customers by entering:

```
SELECT SLSRNUMB
    FROM CUSTOMER
    WHERE BALANCE > CREDLIM
```

and then somehow ask for the number and name of all sales reps whose numbers are in this collection. Actually, these two can be accomplished in a single step by using a feature of SQL called subqueries, as follows:

```
SELECT SLSRNUMB, SLSRNAME
    FROM SLSREP
    WHERE SLSRNUMB IN
        (SELECT SLSRNUMB
            FROM CUSTOMER
            WHERE BALANCE > CREDLIM)
```

In algebra, expressions within parentheses are evaluated first. The same holds true in SQL. The expression within parentheses, which is called a **subquery**, is evaluated first, producing a collection (actually a table) of those sales rep numbers that appear on some row in the *CUSTOMER* table on which the balance is greater than the credit limit. Once this has been accomplished, the remainder of the query can be executed, producing a list of sales rep numbers and names from those rows in the *SLSREP* table for which the sales rep number is in this collection.

**Joining
Multiple
Tables**

Example 15: Joining more than two tables.

Statement: For each part that is on order, list the part number, the number ordered, the order number, the number and name of the customer who placed the order, and the name of the sales rep who represents the customer.

A part is on order if it occurs on any row in the *ORDLNE* table. The part number, number ordered, and order number are all found within the *ORDLNE* table. If these were the only things required for the query, we could just enter:

```
SELECT PARTNUMB, NUMBORD, ORDNUMB
    FROM ORDLNE
```

This is not all we need, however. We also need the customer number, which is in the *ORDERS* table; the customer name, which is in the *CUSTOMER* table; and the sales rep name, which is in the *SLSREP* table. Thus, we really need to join *four* tables: *ORDLNE, ORDERS, CUSTOMER,* and *SLSREP.* The mechanism for doing this is essentially the same as the mechanism for joining two tables. The only difference is that the condition in the WHERE clause will be a compound condition. The WHERE clause in this case would be the following.

```
WHERE ORDERS.ORDNUMB = ORDLNE.ORDNUMB
AND   CUSTOMER.CUSTNUMB = ORDERS.CUSTNUMB
AND   SLSREP.SLSRNUMB = CUSTOMER.SLSRNUMB
```

The first condition relates an order line to an order with a matching order number. The second condition relates this order to the customer with a matching customer number. The final condition relates this customer to a sales rep based on a matching sales rep number.

For the complete query, we list all of the desired columns after the SELECT, qualifying any that appear in more than one table. After the FROM, we list all four tables that are involved in the query. The complete formulation is thus:

```
SELECT ORDLNE.PARTNUMB, NUMBORD, ORDLNE.ORDNUMB, ORDERS.CUSTNUMB,
    CUSTNAME, SLSRNAME
    FROM ORDLNE, ORDERS, CUSTOMER, SLSREP
    WHERE ORDERS.ORDNUMB = ORDLNE.ORDNUMB
    AND   CUSTOMER.CUSTNUMB = ORDERS.CUSTNUMB
    AND   SLSREP.SLSRNUMB = CUSTOMER.SLSRNUMB
```

Result:

PARTNUMB	NUMBORD	ORDNUMB	CUSTNUMB	CUSTNAME	SLSRNAME
AX12	11	12489	124	SALLY ADAMS	MARY JONES
AZ52	2	12498	522	MARY NELSON	SAM BROWN
BA74	4	12498	522	MARY NELSON	SAM BROWN
BT04	1	12491	311	DON CHARLES	SAM BROWN
BT04	1	12500	124	SALLY ADAMS	MARY JONES
BZ66	1	12491	311	DON CHARLES	SAM BROWN
CB03	4	12494	315	TOM DANIELS	WILLIAM SMITH
CX11	2	12495	256	ANN SAMUELS	WILLIAM SMITH
CZ81	2	12504	522	MARY NELSON	SAM BROWN

Question: Could *CUSTOMER.CUSTNUMB* be used in place of *ORDERS.CUSTNUMB* in the query?

Answer: Yes, since the values for these two columns must match by virtue of the condition *CUSTOMER.CUSTNUMB = ORDERS.CUSTNUMB*. Thus, we could choose either one.

Question: Why didn't we have to say *ORDLNE.PARTNUMB*, since *PARTNUMB* also appears as a column in the *PART* table?

Answer: If the *PART* table were used as one of the tables in the query, we would have to qualify *PARTNUMB*. Since it is not, such qualification is unnecessary. (Among the tables listed in the query, only one column is labeled *PARTNUMB*.)

Certainly this last query is more involved than the previous ones. You may be thinking that SQL is not such an easy language to use after all. If you take it one step at a time, however, the query really isn't all that bad. To construct it in a step-by-step fashion we should do the following:

1. List all of the columns that we want printed on the report after the word SELECT. If the name of any column appears in more than one table, precede the column name with the table name (that is, qualify it).
2. List all of the tables involved in the query after the word FROM. This will usually be the tables that contain the columns listed in the SELECT clause. Occasionally, however, there might be a table that does not contain any columns used in the SELECT clause but does contain columns used in the WHERE clause. It must also be listed. (In the last query, if there had been no need to list a customer number or name, but we had needed to list the sales rep name, no columns from the *CUSTOMER* table would be used in the SELECT clause. The *CUSTOMER* table would still have been required, however, since columns from it must have been used in the WHERE clause.)
3. Taking the tables involved one pair at a time, put the condition that relates the tables in the WHERE clause. Join these conditions with AND. If there are any other conditions, they should also be included in the WHERE clause and should be connected to the others with the word AND. If, in the last example, we had only wanted parts present on orders placed by customers with a $500 credit limit, for example, one more condition would have been added to the WHERE clause, giving:

```
SELECT ORDLNE.PARTNUMB, NUMBORD, ORDLNE.ORDNUMB, ORDERS.CUSTNUMB,
    CUSTNAME, SLSRNAME
    FROM ORDLNE, ORDERS, CUSTOMER, SLSREP
    WHERE ORDERS.ORDNUMB = ORDLNE.ORDNUMB
    AND    CUSTOMER.CUSTNUMB = ORDERS.CUSTNUMB
    AND    SLSREP.SLSRNUMB = CUSTOMER.SLSRNUMB
    AND    CREDLIM = 500
```

Update

The remainder of the SQL examples involve the update features of SQL.

Example 16: Changing existing data in the database.

Statement: Change the name of customer 256 TO ANN JONES.

The SQL command to make changes in existing data is the UPDATE command. For this example, the formulation would be:

```
UPDATE CUSTOMER
    SET CUSTNAME = 'ANN JONES'
    WHERE CUSTNUMB = 256
```

Example 17: Adding new data to the database.

Statement: Add sales rep 14, (Name: ANN CRANE, Address: 123 RIVER,ADA,MI, Total Commission: 0, Commission rate: 0.05) to the database.

Addition of new data is accomplished through the INSERT command. If we have specific data, as in this example, we can use the INSERT command as follows:

```
INSERT INTO SLSREP
    VALUES
    (14,'ANN CRANE','123 RIVER,ADA,MI',0.00,0.05)
```

Example 18: Deleting data from the database.

Statement: Delete from the database the customer whose name is AL WILLIAMS.

To delete data from the database, the DELETE command is used, as in the following:

```
DELETE CUSTOMER
    WHERE CUSTNAME = 'AL WILLIAMS'
```

Note that this type of deletion can be very dangerous. If there happened to be another customer with the name AL WILLIAMS, this customer would also be deleted in the process. The safest type of deletion occurs when the condition involves the customer number. In such a case, since the customer number is unique, we can be certain that we will not cause accidental deletion of other rows in the table.

Example 19: Changing data in the database on the basis of a compound condition.

Statement: For each customer with a $500 credit limit whose balance does not exceed his or her credit limit, increase the credit limit to $800.

The only difference between this and the previous update example is that here the condition is compound. Thus, the formulation would be:

```
UPDATE CUSTOMER
    SET CREDLIM = 800
    WHERE CREDLIM = 500
    AND BALANCE < CREDLIM
```

Example 20: Creating a new table with data from an existing table.

Statement: Create a new table called *SMALLCUST* that contains the same columns as *CUSTOMER* but only the rows for which the credit limit is $500 or less.

The first thing we need to do is to describe this new table by means of the data definition facilities of SQL, as follows:

```
CREATE TABLE SMALLCUST
    (CUSTNUMB        DECIMAL(4),
    CUSTNAME         CHAR(15),
    ADDRESS          CHAR(25),
    BALANCE          DECIMAL(7,2),
    CREDLIM          DECIMAL(4),
    SLSRNUMB         DECIMAL(2))
```

Once this is done, we can use the same INSERT command that we encountered in Example 17. Here, however, we use a SELECT command to indicate what is to be inserted into this new table. The exact formulation is:

```
INSERT INTO SMALLCUST
    SELECT *
    FROM CUSTOMER
    WHERE CREDLIM <= 500
```

It should be clear from these examples that SQL is a very powerful language that allows comprehensive queries to be satisfied with brief formulations. The use of SQL is widespread. It is employed with many relational DBMSs, on everything from the largest mainframe to some of the smaller microcomputers, so it's a good idea to become familiar with it.

For other examples of SQL queries, see [8], [9], [11], and [14].

3.3   QBE

In this section, we will investigate a visual approach to manipulating relational databases. It is called **Query-by-Example** (or **QBE** for short) and was proposed by M. M. Zloof (see [16]). It is intended for use on a visual display terminal. Not only are results displayed on the screen in tabular form, but users enter their requests by filling in portions of the displayed tables.

In using QBE, we are first presented with a blank form on the screen:

We indicate which table we wish to manipulate by typing the name of the table in the first box:

ORDERS					

At this point, we can fill in the column headings for those columns we wish to have included in our queries. If we wish to include all columns from the table, we can employ a shortcut. We merely type:

ORDERS P.					

The **P.** stands for print. The system will respond with:

ORDERS	ORDNUMB	ORDDTE	CUSTNUMB		

In the following examples, we will give the QBE formulation but leave it to you to determine the results that the computer would produce. For the first few examples, let's assume that we have requested the part table with the following columns:

PART	PARTNUMB	PARTDESC	UNONHND	ITEMCLSS	WREHSENM

Example 1: Retrieving certain columns and all rows.

Statement: List the part number and description of all parts.

Using the same **P.** that we encountered earlier, we request all part numbers to be printed by putting a **P.** in the *PARTNUMB* and *PARTDESC* columns.

PART	PARTNUMB	PARTDESC	UNONHND	ITEMCLSS	WREHSENM
	P.	P.			

The system would then respond by filling in the part number column with all of the part numbers currently on file. If there are more part numbers than will fit on the screen, which will probably be the case, they will be displayed one screen at a time.

Example 2: Retrieving all columns and all rows.

Statement: List the complete part table.

We could certainly put a **P.** in each column in the table to obtain the desired result. A simpler method is available, however, and that is to put a single **P.** in the first column:

PART	PARTNUMB	PARTDESC	UNONHND	ITEMCLSS	WREHSENM
P.					

This indicates that we want the full table printed.

Example 3: Retrieval with a simple condition.

Statement: List the part numbers of all parts in item class HW.

We use the **P.**, as before, to indicate the columns that are to be printed. We can also place a specific value in a column:

PART	PARTNUMB	PARTDESC	UNONHND	ITEMCLSS	WREHSENM
	P.			HW	

This indicates that the part numbers to be printed should only be those for which the item class is HW.

Example 4: Retrieval with a compound condition involving AND.

Statement: List the part numbers for all parts which are in item class HW and are located in warehouse 3.

As you might expect, we can put specific values in more than one column. Further, we can also use the comparison operators =, >, > =, <, and < =, as well as ~ = (NOT EQUAL). It is common in QBE to omit the = symbol in "equal" and "not equal" comparisons, although it may be used if desired.

PART	PARTNUMB	PARTDESC	UNONHND	ITEMCLSS	WREHSENM
	P.			HW	3

In this case, we are requesting those parts for which the item class is HW *and* the warehouse is 3.

Example 5: Retrieval with a compound condition involving OR.

Statement: List the part numbers for those parts that are in item class HW or warehouse 3.

What we essentially have in this query is two queries. We want all parts that are in class HW. We also want all parts that are in warehouse 3. This is effectively how we enter our request, as two queries:

PART	PARTNUMB	PARTDESC	UNONHND	ITEMCLSS	WREHSENM
	P.			HW	
	P.				3

The first row indicates that we want all parts that are in class HW. The second row indicates that we also want all parts that are in warehouse 3.

Example 6: Retrieval using NOT.

Statement: List the part numbers of all parts that are not in item class HW.

We use the symbol ~ for NOT and enter:

PART	PARTNUMB	PARTDESC	UNONHND	ITEMCLSS	WREHSENM
	P.			~HW	

In each of the prior examples, we could have used a feature of the QBE language from which it draws its name: an example. The prior queries were simple enough, so there was no real need to use an example, but it would certainly have been legitimate to do so. To use an example, we pick a sample response that the computer could give to the query and enter it in the table. To indicate that it is merely an example, we underline it. Thus, in the previous query, we could have entered:

PART	PARTNUMB	PARTDESC	UNONHND	ITEMCLSS	WREHSENM
	P.XYZ			~HW	

We are indicating that XYZ is an example of the response we are expecting. It does *not* have to be an actual response that would be generated in response to this query. In this case, in fact, there is not even a part XYZ in the database. Notice the difference between the part number XYZ and the item class HW. The part number is underlined, indicating that it is strictly an example. The item class is not underlined, indicating that it is an actual value.

Working with a single table is thus fairly simple. But what if we need to use two tables and the relationship between them in order to obtain the desired information? This is also possible in QBE. To illustrate the method, let's assume that we will use the following two tables.

CUSTOMER	CUSTNUMB	CUSTNAME	BALANCE	CREDLIM	SLSRNUMB

and

SLSREP	SLSRNUMB	SLSRNAME	SLSRADDR	TOTCOMM	COMMRATE

Example 7: Retrieval using more than one table.

Statement: For each customer whose credit limit is $800, list the number and name of the customer together with the number and name of the sales rep who represents the customer.

This query cannot be satisfied by using a single table. The customer name is in the customer table, whereas the sales rep name is in the sales rep table. We need the equivalent of a **join** operation. This is accomplished by having two tables on the screen and filling them in as follows:

CUSTOMER	CUSTNUMB	CUSTNAME	BALANCE	CREDLIM	SLSRNUMB
	P.	P.		800	123

SLSREP	SLSRNUMB	SLSRNAME	SLSRADDR	TOTCOMM	COMMRATE
	P.123	P.			

In this case, there is a **P.** in both the *CUSTNUMB* and *CUSTNAME* columns of the *CUSTOMER* table, indicating that these are the columns in which results are to be printed. Further, there is an example, 123, in the sales rep number column of the *CUSTOMER* table, as well as *the same* example in the sales rep number column of the *SLSREP* table. These examples are necessary, and the fact that they are the same is *crucial*. This is what tells the system how the tables are to be joined. Finally, since the 800 is not underlined, it is *not* an example but rather a specific restriction. In other words, we are telling the system to PRINT the number and name of any customers in the

CUSTOMER table for whom the value in the credit limit column is 800 and for whom there is a row in the SLSREP table where the sales rep number matches this sales rep number in the CUSTOMER table.

For other examples of QBE queries, see [8], [9], [11], and [14].

3.4   THE RELATIONAL ALGEBRA

The relational algebra is one of the original approaches proposed for manipulating relational databases. In the relational algebra, operations act on tables to produce new tables, just as the operations of +, -, and so on, act on numbers to produce new numbers in the algebra with which you are familiar. Retrieving data from a relational database through the use of the relational algebra involves using these operations on existing tables to form a new table that contains the desired information. It may be necessary to use successive commands to form intermediate tables before the final result is obtained, as some of the following examples demonstrate. As you will notice in these examples, each command ends with a clause that reads GIVING, followed by a table name. This clause is requesting that the result of the execution of the command is to be placed in a table with the name we have specified.

There are eight commands that make up the relational algebra. In this text, we will examine three of the most important of them: SELECT, PROJECT, and JOIN.

Select

The relational algebra SELECT command takes a horizontal subset of a table, that is, it causes only certain rows to be included in the new table. (It should not be confused with the SQL SELECT command, which actually includes the power to accomplish all of the relational algebra commands.) This SELECT causes the creation of a new table that consists of only those rows from the indicated table which meet some specified criteria.

Statement: List all information from the CUSTOMER table concerning customer 256.

In relational algebra:

```
SELECT CUSTOMER WHERE CUSTNUMB = 256 GIVING ANSWER.
```

In SQL:

```
SELECT * FROM CUSTOMER WHERE CUSTNUMB = 256
```

Project

The PROJECT command within the relational algebra takes a vertical subset of a table, that is, it causes only certain columns to be included in the new table.

Statement: List the number and name of all customers.

In relational algebra:

```
PROJECT CUSTOMER OVER (CUSTNUMB, CUSTNAME) GIVING ANSWER
```

In SQL:

```
SELECT CUSTNUMB, CUSTNAME FROM CUSTOMER
```

We can combine SELECT and PROJECT, as the following example shows:

Statement: List the number and name of all customers who have an $800 credit limit.

This is accomplished in a two-step process. We first use a SELECT command to create a new table that contains only those customers who have the appropriate credit limit. Then we project that table to restrict the result to only the indicated columns.

In relational algebra:

```
SELECT CUSTOMER WHERE CREDLIM = 800 GIVING TEMP
PROJECT TEMP OVER (CUSTNUMB, CUSTNAME) GIVING ANSWER
```

In SQL:

```
SELECT CUSTNUMB, CUSTNAME FROM CUSTOMER
    WHERE CREDLIM = 800
```

Join

The JOIN operation is at the heart of the relational algebra. It is the command that allows us to pull together data from more than one table. In the most usual form of the JOIN, two tables are **join**ed together on the basis of a common attribute. A new table is formed, containing the columns of both the tables that have been joined. Rows in this new table will be the concatenation (combination) of a row from the first table and a row from the second that match on the common attribute (often called the JOIN column). For example, suppose we wanted to JOIN the following tables, *CUSTOMER* and *SLSREP*, on *SLSRNUMB* (the join column) as follows.

CUSTOMER

CUSTNUMB	CUSTNAME	ADDRESS	SLSRNUMB
124	SALLY ADAMS	481 OAK,LANSING,MI	3
256	ANN SAMUELS	215 PETE,GRANT,MI	6
311	DON CHARLES	48 COLLEGE,IRA,MI	12
315	TOM DANIELS	914 CHERRY,KENT,MI	6
405	AL WILLIAMS	519 WATSON,GRANT,MI	12
412	SALLY ADAMS	16 ELM,LANSING,MI	3
522	MARY NELSON	108 PINE,ADA,MI	12
567	JOE BAKER	808 RIDGE,HARPER,MI	6
587	JUDY ROBERTS	512 PINE,ADA,MI	6
622	DAN MARTIN	419 CHIP,GRANT,MI	3

SLSREP

SLSRNUMB	SLSRNAME
3	MARY JONES
6	WILLIAM SMITH
12	SAM BROWN

Suppose that we wish to call the result of this join *TEMP*. It would look like:

TEMP

CUSTNUMB	CUSTNAME	ADDRESS	SLSRNUMB	SLSRNAME
124	SALLY ADAMS	481 OAK,LANSING,MI	3	MARY JONES
256	ANN SAMUELS	215 PETE,GRANT,MI	6	WILLIAM SMITH
311	DON CHARLES	48 COLLEGE,IRA,MI	12	SAM BROWN
315	TOM DANIELS	914 CHERRY,KENT,MI	6	WILLIAM SMITH
405	AL WILLIAMS	519 WATSON,GRANT,MI	12	SAM BROWN
412	SALLY ADAMS	16 ELM,LANSING,MI	3	MARY JONES
522	MARY NELSON	108 PINE,ADA,MI	12	SAM BROWN
567	JOE BAKER	808 RIDGE,HARPER,MI	6	WILLIAM SMITH
587	JUDY ROBERTS	512 PINE,ADA,MI	6	WILLIAM SMITH
622	DAN MARTIN	419 CHIP,GRANT,MI	3	MARY JONES

Note that the column on which the tables are joined appears only once. Other than that, all columns from both tables are present in the result.

The output from the join can be restricted to include only desired columns by using the PROJECT command, as the following example illustrates.

Statement: List the number and name of all customers together with the number and name of the sales rep who represents each customer.

In relational algebra:

```
JOIN CUSTOMER SLSREP
     WHERE CUSTOMER.SLSRNUMB = SLSREP.SLSRNUMB
     GIVING TEMP
PROJECT TEMP OVER (CUSTNUMB, CUSTNAME, SLSRNUMB,
     SLSRNAME) GIVING ANSWER
```

In SQL:

```
SELECT CUSTNUMB, CUSTNAME, SLSRNUMB, SLSRNAME
     FROM CUSTOMER, SLSREP
     WHERE CUSTOMER.SLSRNUMB = SLSREP.SLSRNUMB
```

This should give you the flavor of the relational algebra. Other types of joins and other commands are present in the relational algebra, but they are beyond the scope of this text. For information on these topics, see [9] and [14].

3.5   NATURAL LANGUAGES

Another approach to manipulating databases for the purpose of retrieval involves the use of so-called **natural languages**. When employing these languages, users type their queries as normal English questions or requests. These languages possess a built-in dictionary of words that they can interpret. Users can add words to this dictionary in order to tailor it to their particular application. In addition, these languages recognize the names of tables and columns in the database. Finally, they can recognize alternate names, or aliases, for table and column names; for example, *account number* may be another name for *customer number*.

These languages are discussed here because they represent an important way of retrieving data from a relational database. (They can be used with databases that follow other models, but for them to reach their full potential, the flexibility of the relational model is crucial.) Let's look at a few examples of querying a database through a fictitious (but representative) natural language called NL; in each case, an SQL formulation for the same query will also be presented.

Statement: Find the name of customer 256.

In NL:

```
User: What is the nme of customer 256?
Computer: I don't recognize "nme."  Did you mean "name"?
User: Yes.
Computer:      CUSTNAME
               ANN SAMUELS
```

In SQL:

```
SELECT CUSTNAME
     FROM CUSTOMER
     WHERE CUSTNUMB = 256
```

Statement: Find the names of all of the customers of sales rep 6.

In NL:

```
User: Give me the names of the customers of sales rep 6.
Computer:      CUSTNAME
               ANN SAMUELS
               TOM DANIELS
               JOE BAKER
               JUDY ROBERTS
```

In SQL:

```
SELECT CUSTNAME
     FROM CUSTOMER
     WHERE SLSRNUMB = 6
```

Statement: Of the names found in the previous query, which ones have a credit limit of $300?

In NL:

```
User: Which ones have a credit limit of 300?
Computer:      CUSTNAME
               TOM DANIELS
               JOE BAKER
```

In SQL:

```
SELECT CUSTNAME
     FROM CUSTOMER
     WHERE SLSRNUMB = 6
     AND CREDLIM = 300
```

One final note: if the user had placed the term *small customer* in the dictionary and had defined it as a customer whose credit limit is $300, then that last query could have been formulated as "Which ones are small customers?"

SUMMARY

1. SQL (Structured Query Language) is a language that is used to manipulate relational databases. The basic form of an SQL command is SELECT-FROM-WHERE.
   a. The CREATE command is used to describe the layout of a table.
   b. Computed columns can be used in SQL commands.
   c. Sorting is accomplished through the use of the ORDER BY clause.
   d. SQL has the built-in functions COUNT, SUM, AVG, MAX, and MIN.
   e. Joining two tables is accomplished in SQL through the use of a condition that relates matching rows in the tables to be joined, for example
      CUSTOMER.SLSRNUMB = SLSREP.SLSRNUMB.
   f. One query nested within another query is called a subquery and is executed first.
   g. The INSERT command is used to add new data.
   h. The UPDATE command is used to change existing data.
   i. The DELETE command is used to delete existing data.
2. QBE (Query-by-Example) is another language that is used to manipulate relational databases. QBE queries are indicated by filling in tables on the screen.
   a. P. is included in the columns to be printed.
   b. Values that are not underlined indicate specific restrictions.
   c. Underlined values are examples. They are optional unless they are used to relate two tables.
3. The relational algebra is an approach to manipulating relational databases in which there are operations that act on tables to produce new tables. The three most important operations are SELECT, PROJECT, and JOIN.
   a. The SELECT operation selects only certain rows from a table.
   b. The PROJECT operation selects only certain columns from a given table.
   c. The JOIN operation combines data from two tables.
4. Natural languages form yet another way of retrieving data from a relational database. Users type their requests in ordinary English when they use natural languages.

KEY TERMS

built-in function	relational algebra
computed column	relational model
join	select
natural language	SQL
project	subquery
Query-by-Example (QBE)	

EXERCISES

Questions 1 through 17 are based on the sample database of Figure 3.1 and deal with the language SQL. For each question, give both the appropriate SQL formulation and the result that would be produced.

1. Find the part number and description of all parts.
2. List the complete sales rep table.

3. Find the names of all the customers who have a credit limit of at least $800.
4. Give the order numbers of those orders placed by customer 124 on September 5, 1994.
5. Give the part number, description, and on-hand value (units on hand * price) for each part in item class AP. (On-hand value is really units on hand * cost but we do not have a cost column in the *PART* table.)
6. Find the number and name of all customers whose last name is NELSON.
7. List all details about parts. The output should be sorted by unit price.
8. Find out how many customers have a balance that exceeds their credit limit.
9. Find the total of the balances for all the customers represented by sales rep 12.
10. For each order, list the order number, the order date, the customer number, and the customer name.
11. For each order placed on September 5, 1994, list the order number, the order date, the customer number, and the customer name.
12. Find the number and name of all sales reps who represent any customer with a credit limit of $1000. Do this in two different ways: in one solution use a subquery; in the other, do not use a subquery.
13. For each order, list the order number, the order date, the customer number, the customer name, together with the number and name of the sales rep who represents the customer.
14. Change the description of part BT04 to OVEN.
15. Add order 12600 (date: 90694, customer number: 311) to the database.
16. Delete all customers whose balance is $0.00 and who are represented by sales rep 12.
17. Describe a new table to the database called *SPGOOD*. It contains only part number, description, and price. Once this has been done, insert the part number description and price of all parts whose item class is SG into this new table.

Questions 18 through 24 are also based on the sample database of Figure 3.1 but deal with QBE. For each question, give the appropriate QBE formulation.

18. List the number and name of all sales reps.
19. List the complete *CUSTOMER* table.
20. List the number and name of all customers represented by sales rep 3.
21. List the number and name of all customers represented by sales rep 3 and whose credit limit is $500.
22. List the number and name of all customers represented by sales rep 3 or who have a credit limit of $500.
23. List the number and name of all customers who are not represented by sales rep 3.
24. List the number and name of all customers who are represented by MARY JONES.

Questions 25 through 28 are also based on the sample database of Figure 3.1 but deal with the relational algebra. For each question, give both the appropriate relational algebra formulation and an equivalent SQL formulation.

25. List all information from the part table concerning part BT04.
26. List the number and name of all sales reps.
27. List the order number, order date, customer number, and customer name for each order.
28. List the order number, order date, customer number, and customer name for each order placed by any customer represented by MARY JONES.

Relational Model II: Advanced Topics

OBJECTIVES

1. Discuss views: what they are, how they are described, and how they are used.
2. Discuss the use of indexes for improving performance.
3. Define the catalog and explain its use.

4. Explain entity and referential integrity.
5. Discuss the manner in which the structure of a relational database can be changed.
6. Describe the characteristics a system must possess in order to be relational.

4.1 INTRODUCTION

In the last chapter, we examined data definition and manipulation within the relational model. In this chapter, we will investigate some other aspects of the model. In section 4.2, we will look at views, which represent a way of giving each user his or her own picture of the database to work with. In section 4.3, we will look at indexes and their use in improving performance. In section 4.4, we will discuss the catalog that is furnished by many relational DBMSs to provide users with access to information about the structure of a database. Two critical integrity rules are presented in section 4.5. Section 4.6 covers one of the real strengths of the relational model, the ease with which a database structure can be changed. Finally, in section 4.7, we will look at a very important question: How can we tell whether a system is really relational? With so many systems claiming in their advertising literature to be "relational," this question is especially significant.

In the discussion that follows, we will use SQL as a mechanism for illustrating the concepts. It should be emphasized, however, that many systems which do not support the SQL language provide the features we will be discussing. Although the manner in which this is accomplished varies slightly from one system to another, the basic concepts are the same, and it should be easy for you to transfer the knowledge you gain in this chapter to a non-SQL system.

4.2 VIEWS

Most relational mainframe DBMSs and some of the microcomputer DBMSs support the concept of a view. A **view** is basically an individual user's picture of the database. In many cases, a user can interact with the database via a view. Since a view is usually much less involved than the full database, its use can represent a great simplification. Views also provide a measure of security, since omitting sensitive tables or columns from a view will render them unavailable to anyone who is accessing the database via the view.

To illustrate the idea of a view, let's suppose that Bill is interested in the part number, description, units on hand, and unit price of those parts which are in item class HW. He is not interested in any of the other columns in the *PART* table. Nor is he interested in any of the rows that correspond to parts in other item classes. Life would certainly be simpler for Bill if the other rows and columns were not even present. While we cannot change the structure of the *PART* table and omit some of its rows just for Bill, we can do the next best thing. We can provide him a view that consists of precisely the rows and columns he is interested in. Using SQL, we do this as follows:

```
CREATE VIEW HOUSEWARES AS
     SELECT PARTNUMB, PARTDESC, UNONHAND,
          UNITPRCE
          FROM PART
          WHERE ITEMCLSS = 'HW'
```

The SELECT command, which is called the **defining query**, indicates precisely what is to be included in the view. Notice that it is exactly what Bill wants. Conceptually, given the current data in the *PREMIERE PRODUCTS* database, this view will contain the data shown in Figure 4.1. The data does not really exist in this form, however, nor will it *ever* exist in this form. It is tempting to think that when this view is used, the query will be executed and will produce some sort of temporary table, called *HOUSEWARES*, which Bill could then access. This is *not* what happens.

Figure 4.1

HOUSEWARES view

HOUSEWARES

PARTNUMB	PARTDESC	UNONHAND	UNITPRCE
AX12	IRON	104	17.95
BH22	TOASTER	95	34.95
CA14	SKILLET	2	19.95
CX11	MIXER	112	57.95

Instead, the query acts as a sort of window into the database (see Figure 4.2). As far as Bill is concerned, the whole database is just the portion shown in dark blue. Any change that affects the dark portion of the *PART* table will be seen by Bill. He will be totally unaware, however, of a change that affects any other part of the database.

Figure 4.2

PREMIERE PRODUCTS sample data

PART

PARTNUMB	PARTDESC	UNONHAND	ITEMCLSS	WREHSENM	UNITPRCE
AX12	IRON	104	HW	3	17.95
AZ52	SKATES	20	SG	2	24.95
BA74	BASEBALL	40	SG	1	4.95
BH22	TOASTER	95	HW	3	34.95
BT04	STOVE	11	AP	2	402.99
BZ66	WASHER	52	AP	3	311.95
CA14	SKILLET	2	HW	3	19.95
CB03	BIKE	44	SG	1	187.50
CX11	MIXER	112	HW	3	57.95
CZ81	WEIGHTS	208	SG	2	108.99

The way in which this is implemented is really clever. Suppose, for example, that Bill were to type the following query:

```
SELECT *
     FROM HOUSEWARES
     WHERE UNONHAND > 100
```

The query would *not* be executed in this form. Instead, it would be merged with the query that defines the view to form the query that is actually executed. In this case, the merging of the two would form:

```
SELECT PARTNUMB, PARTDESC, UNONHAND,
     UNITPRCE
     FROM PART
     WHERE ITEMCLSS = 'HW'
          AND UNONHAND > 100
```

Notice the following three things: the selection is from the *PART* table rather than the *HOUSEWARES* view; the * is replaced by those columns which are in the *HOUSE-WARES* view; and the condition involves the condition in the query entered by Bill together with the condition stated in the view definition. This new query is the one that is actually executed.

Bill, however, is unaware that this kind of activity is taking place. It seems to him that there really is a table called *HOUSEWARES* that is being accessed. One advantage of this approach is that since *HOUSEWARES* never exists in its own right, any update to the *PART* table is *immediately* available to someone accessing the database through the view. If *HOUSEWARES* were an actual stored table, this would not be the case.

What if Bill wanted different names for the columns? This could be accomplished by including the desired names in the CREATE VIEW statement. For example, if Bill wanted the names of the part number, description, units on hand, and price columns to be *PNUM, DESC, ON_HAND,* and *PRICE*, respectively, the CREATE VIEW statement would be:

```
CREATE VIEW HOUSEWARES (PNUM, DESC, ON_HAND, PRICE) AS
     SELECT PARTNUMB, PARTDESC, UNONHAND,
          UNITPRCE
          FROM PART
          WHERE ITEMCLSS = 'HW'
```

In this case, when Bill accessed the *HOUSEWARES* view, he would refer to *PNUM, DESC, ON_HAND,* and *PRICE* rather than *PARTNUMB, PARTDESC, UNONHAND,* and *UNITPRCE*.

The *HOUSEWARES* view is an example of a row-and-column subset view; that is, it consists of a subset of the rows and columns in some individual table, in this case the *PART* table. Since the query can be any SQL query, a view could also involve the join of two or more tables.

Suppose, for example, that Joan needed to know the number and name of each sales rep, along with the number and name of the customers represented by each sales rep. It would be much simpler for her if this information were in a single table instead of two tables that had to be joined together. She would really like a single table that contained a sales rep number, sales rep name, customer number, and customer name. Suppose she would also like these columns to be named *SNUM, SNAME, CNUM,* and *CNAME*, respectively. This could be accomplished by using a join in the CREATE VIEW statement, as follows:

```
CREATE VIEW SLSCUST (SNUMB, SNAME, CNUMB, CNAME) AS
     SELECT SLSREP.SLSRNUMB, SLSREP.SLSRNAME,
          CUSTOMER.CUSTNUMB, CUSTOMER.NAME
          FROM SLSREP, CUSTOMER
          WHERE SLSREP.SLSRNUMB =
               CUSTOMER.SLSRNUMB
```

Given the current data in the *PREMIERE PRODUCTS* database, this view is conceptually the table shown in Figure 4.3.

Figure 4.3

SLSCUST view

SLSCUST

SNUMB	SNAME	CNUMB	CNAME
3	MARY JONES	124	SALLY ADAMS
3	MARY JONES	412	SALLY ADAMS
3	MARY JONES	622	DAN MARTIN
6	WILLIAM SMITH	256	ANN SAMUELS
6	WILLIAM SMITH	315	TOM DANIELS
6	WILLIAM SMITH	567	JOE BAKER
6	WILLIAM SMITH	587	JUDY ROBERTS
12	SAM BROWN	311	DON CHARLES
12	SAM BROWN	405	AL WILLIAMS
12	SAM BROWN	522	MARY NELSON

As far as Joan is concerned, this is a real table; she does not need to know what goes on behind the scenes in order to use it. She could find the number and name of the sales rep who represents customer 256, for example, merely by entering:

```
SELECT SNUMB, SNAME
     FROM SLSCUST
     WHERE CNUMB = 256
```

She will be completely unaware that, behind the scenes, her query is actually converted to:

```
SELECT SLSREP.SLSRNUMB, SLSREP.SLSRNAME
     FROM SLSREP, CUSTOMER
     WHERE SLSREP.SLSRNUMB =
          CUSTOMER.SLSRNUMB
          AND CUSTNUMB = 256
```

As you can see, the use of views can greatly simplify the process of querying a database. In general, the more complicated the defining query, the more simplification the user is provided by the view. Not everything here is rosy, however. Problems arise when one tries to update a database through a view, especially a view whose defining query involves a join. This problem and others like it are beyond the scope of this text; for information about them, see [9] and [14]. Nevertheless, views are a helpful feature from which substantial benefits can be achieved.

4.3 INDEXES

If you wanted to find a discussion of a given topic in a book, you could scan the entire book from start to finish, looking for references to the topic you had in mind. More than likely, however, you wouldn't have to resort to this technique. If the book had a good index, you could use it to rapidly locate the pages on which your topic was discussed.

Within relational model systems on both mainframes and microcomputers, the main mechanism for increasing the efficiency with which data is retrieved from the database is the use of **indexes**. Conceptually, these indexes are very much like the index in a book. Consider Figure 4.4, for example, which shows the *CUSTOMER* table for *PREMIERE PRODUCTS* together with one extra column, *REC*. This extra column gives the number of each record within the file. (Customer 124 is on record one; customer 256 is on record two; and so on.) These record numbers are used by the DBMS, not by the users, and that is why we do not normally show them. Here, however, we are dealing with the manner in which the DBMS works, so we do need to be aware of them.

Figure 4.4

CUSTOMER table with record numbers

CUSTOMER

REC	CUSTNUMB	CUSTNAME	ADDRESS	BALANCE	CREDLIM	SLSRNUMB
1	124	SALLY ADAMS	481 OAK,LANSING,MI	418.75	500	3
2	256	ANN SAMUELS	215 PETE,GRANT,MI	10.75	800	6
3	311	DON CHARLES	48 COLLEGE,IRA,MI	200.10	300	12
4	315	TOM DANIELS	914 CHERRY,KENT,MI	320.75	300	6
5	405	AL WILLIAMS	519 WATSON,GRANT,MI	201.75	800	12
6	412	SALLY ADAMS	16 ELM,LANSING,MI	908.75	1000	3
7	522	MARY NELSON	108 PINE,ADA,MI	49.50	800	12
8	567	JOE BAKER	808 RIDGE,HARPER,MI	201.20	300	6
9	587	JUDY ROBERTS	512 PINE,ADA,MI	57.75	500	6
10	622	DAN MARTIN	419 CHIP,GRANT,MI	575.50	500	3

Figure 4.5

Index for *CUSTOMER* table on *CUSTNUMB* column

In order to rapidly access a customer on the basis of his or her number, we might choose to create and use an index as shown in Figure 4.5. The index has two columns.

CUSTNUMB INDEX

CUSTNUMB	REC
124	1
256	2
311	3
315	4
405	5
412	6
522	7
567	8
587	9
622	10

CUSTOMER

REC	CUSTNUMB	CUSTNAME	ADDRESS	BALANCE	CREDLIM	SLSRNUMB
1	124	SALLY ADAMS	481 OAK,LANSING,MI	418.75	500	3
2	256	ANN SAMUELS	215 PETE,GRANT,MI	10.75	800	6
3	311	DON CHARLES	48 COLLEGE,IRA,MI	200.10	300	12
4	315	TOM DANIELS	914 CHERRY,KENT,MI	320.75	300	6
5	405	AL WILLIAMS	519 WATSON,GRANT,MI	201.75	800	12
6	412	SALLY ADAMS	16 ELM,LANSING,MI	908.75	1000	3
7	522	MARY NELSON	108 PINE,ADA,MI	49.50	800	12
8	567	JOE BAKER	808 RIDGE,HARPER,MI	201.20	300	6
9	587	JUDY ROBERTS	512 PINE,ADA,MI	57.75	500	6
10	622	DAN MARTIN	419 CHIP,GRANT,MI	575.50	500	3

The first column contains a customer number, and the second column contains the number of the record on which the customer is found. Since customer numbers are unique, there will be only a single record number. This need not always be the case, however. Suppose, for example, that we wanted to rapidly access all customers who had a given credit limit or all customers who were represented by a given sales rep. We might choose to create and use an index on credit limit as well as an index on sales rep number. These two indexes, along with the index on the customer number are shown in Figure 4.6. In the index on credit limit, the first column contains a credit limit, and the second column contains the numbers of *all* the records on which that credit limit is found. The index on sales rep number is similar. (Actually, the structure used for an index is usually a little more complicated than these examples show. They are perfectly acceptable for our purposes, however. For more information about the structure of these indexes, see [14].)

Figure 4.6

Three indexes for
CUSTOMER table

CUSTNUMB INDEX	
CUSTNUMB	REC
124	1
256	2
311	3
315	4
405	5
412	6
522	7
567	8
587	9
622	10

CREDLIM INDEX	
CREDLIM	RECs
300	3, 4, 8
500	1, 9, 10
800	2, 5, 7
1000	6

SLSRNUMB INDEX	
SLSRNUMB	RECs
3	1, 6, 10
6	2, 4, 8, 9
12	3, 5, 7

CUSTOMER

REC	CUSTNUMB	CUSTNAME	ADDRESS	BALANCE	CREDLIM	SLSRNUMB
1	124	SALLY ADAMS	481 OAK,LANSING,MI	418.75	500	3
2	256	ANN SAMUELS	215 PETE,GRANT,MI	10.75	800	6
3	311	DON CHARLES	48 COLLEGE,IRA,MI	200.10	300	12
4	315	TOM DANIELS	914 CHERRY,KENT,MI	320.75	300	6
5	405	AL WILLIAMS	519 WATSON,GRANT,MI	201.75	800	12
6	412	SALLY ADAMS	16 ELM,LANSING,MI	908.75	1000	3
7	522	MARY NELSON	108 PINE,ADA,MI	49.50	800	12
8	567	JOE BAKER	808 RIDGE,HARPER,MI	201.20	300	6
9	587	JUDY ROBERTS	512 PINE,ADA,MI	57.75	500	6
10	622	DAN MARTIN	419 CHIP,GRANT,MI	575.50	500	3

Typically, an index can be created and maintained for any column or combination of columns in any table. Once an index has been created, it can be used to facilitate retrieval. In powerful mainframe relational systems, the decision concerning which index or indexes to use (if any) during a particular type of retrieval is one function of a part of the DBMS called an **optimizer**. (No reference is made to any index by the user; rather, the system makes the decision behind the scenes.) In less powerful systems and, in particular, in many of the microcomputer systems, the user may have to specifically indicate in some fashion that a given index should be used.

As you would expect, the use of any index is not purely advantageous or disadvantageous. The advantage was already mentioned: an index makes certain types of retrieval more efficient. There are two disadvantages. First, an index occupies space that could be used for something else. Any retrieval that can be made using an index can also be made without the index. The process may be less efficient, but it is still possible. So an index, while it occupies space, is technically not necessary. The other disadvantage is that the index must be updated whenever corresponding data in the database is updated. Without the index, these updates would not have to be performed. The main question that we must ask when considering whether or not to create a given index is: Do the benefits derived during retrieval outweigh the additional storage required and the extra processing involved in update operations?

Indexes can be added and dropped at will. The final decision concerning the columns or combination of columns on which indexes should be built does not have to be made at the time the database is first implemented. If the pattern of access to the database later indicates that overall performance would benefit from the creation of a new index, it can easily be added. Likewise, if it appears that an existing index is unnecessary, it can easily be dropped.

For further information concerning indexes, see [9] and [14].

4.4   THE CATALOG

Information about tables that are known to the system is kept in the system **catalog**. This section will describe the types of things kept in a catalog and the way the catalog can be queried to determine information about the database structure. (This description happens to represent the way things are done in DB2, IBM's mainframe relational DBMS.) Although catalogs in individual relational DBMSs will vary from what is shown here, the general ideas apply to most relational systems.

The catalog we will look at contains three tables; *SYSTABLES* (information about the tables known to SQL), *SYSCOLUMNS* (information about the columns within these tables), and *SYSINDEXES* (information about the indexes that are defined on these tables). While these tables would have many columns, only a few are of concern to us here.

SYSTABLES contains columns *NAME*, *CREATOR*, and *COLCOUNT*. The *NAME* column identifies the name of a table. The *CREATOR* column contains an identification of the person or group that created the table. The *COLCOUNT* column contains the number of columns within the table that is being described. If, for example, the user whose ID is SALESX01 created the sales rep table and the sales rep table had five columns, there would be a row in the *SYSTABLES* table in which *NAME* was SLSREP, *CREATOR* was SALESX01, and *COLCOUNT* was 5. Similar rows would exist for all tables known to the system.

SYSCOLUMNS contains columns *NAME*, *TBNAME*, and *COLTYPE*. The *NAME* column identifies the name of a column in one of the tables. The table in which the column is found is stored in *TBNAME*, and the data type for the column is found in *COLTYPE*. There will be a row in *SYSCOLUMNS* for each column in the *SLSREP* table, for example. On each of these rows, *TBNAME* will be SLSREP. On one of these rows, *NAME* will be SLSRNUMB and *COLTYPE* will be DECIMAL(2). On another row, *NAME* will be SLSRNAME and *COLTYPE* will be CHAR(15).

SYSINDEXES contains columns *NAME*, *TBNAME*, and *CREATOR*. The name of the index is found in the *NAME* column. The name of the table on which the index was built is found in the *TBNAME* column. The ID of the person or group that created the index is found in the *CREATOR* column.

The system catalog is a relational database of its own. Consequently, in general, the same types of queries that are used to retrieve information from relational databases can be used to retrieve information from the system catalog. The following queries illustrate this process.

1. List the name and creator of all tables known to the system.

```
SELECT NAME, CREATOR
    FROM SYSTABLES
```

2. List all of the columns in the *CUSTOMER* table as well as their associated data types.

```
SELECT NAME, COLTYPE
    FROM SYSCOLUMNS
    WHERE TBNAME = 'CUSTOMER'
```

3. List all tables that contain a column called *SLSRNUMB*.

```
SELECT TBNAME
    FROM SYSCOLUMNS
    WHERE NAME = 'SLSRNUMB'
```

Thus, information about the tables that are in place in our relational database, the columns they contain, and the indexes built on them can be obtained from the catalog by using the same SQL syntax that is used to query any other relational database. We don't need to worry about updating these tables; the system will do it for us automatically every time a change is made in the database structure.

4.5 INTEGRITY RULES

There are two very special rules that should be enforced by a relational DBMS. They were defined by Codd in [6], and they relate to two special types of keys: **primary keys** and **foreign keys**. The two integrity rules are called **entity integrity** and **referential integrity**.

Entity Integrity

In some DBMSs, when we describe a database, we can indicate that certain columns can accept a special value called **null**. Essentially, setting the value in a given column to null is like not filling it in at all. It is used when a value is unknown or inapplicable. It is *not* the same as blank or zero, which are actual values. For example, the value of zero in *BALANCE* indicates that the customer has a zero balance. A value of null, on the other hand, indicates that, for whatever reason, the customer's balance is unknown.

If we indicate that the column *BALANCE* can be null, we are saying that this situation (a customer with an unknown balance) is something we want to allow. If we don't want to allow it, we indicate that *BALANCE* cannot be null.

The decision as to whether to allow nulls is generally made on a column-by-column basis. There is one type of column for which we should *never* allow nulls, however, and that is the **primary key**. After all, the primary key is supposed to uniquely identify a given row, and this could not happen if nulls were allowed. How, for example, could we tell two customers apart if both had a null customer number? The restriction that the primary key cannot allow null values is called entity integrity.

> *Definition:* **Entity integrity** is the rule that no column that participates in the primary key may accept null values.

This property guarantees that each entity will indeed have its own identity. In other words, preventing the primary key from accepting null values ensures that one entity can be distinguished from another.

Referential Integrity

In the relational model as we've been discussing it up until now, relationships are not explicit. They are accomplished by having common columns in two or more tables. The one-to-many relationship between sales reps and customers, for example, is accomplished by including *SLSRNUMB*, the primary key of the *SLSREP* table, as a column in the *CUSTOMER* table.

This approach has its problems. First of all, relationships are not very obvious. If we were not already familiar with the relationships within the *PREMIERE PRODUCTS* database, we would have to note the matching columns in separate tables in order to be aware of a relationship. Even then, we couldn't be sure. Two columns having the same name could be just a coincidence. These columns might have nothing to do with each other. Second, what if the key to the *SLSREP* table were *SLSRNUMB* but the corresponding column within the *CUSTOMER* table happened to be called *SLSRNUMB*? Unless we were aware that these two columns were really the same, the relationship between customers and sales reps would not be clear. In a database having as few tables and columns as the *PREMIERE PRODUCTS* database, these problems might not be major ones. But picture a database that has twenty tables, each one containing an average of thirty columns. As the number of tables and columns increases, so do the problems.

There is also another problem. Nothing about the model itself would prevent a user from storing a customer whose sales rep number did not correspond to any sales rep already in the database. Clearly this is not a desirable situation.

Fortunately, a solution has been found for these two problems, and involves the use of foreign keys. A **foreign key** is a column (or collection of columns) in one table whose value is required to match the value of the primary key for some row in another table. The *SLSRNUMB* in the *CUSTOMER* table is a foreign key that must match the primary key of the *SLSREP* table. In practice, this simply means that the sales rep number for any customer must be the same as the number of some sales rep who is already in the database.

There is one possible exception to this. Some organizations do not require a customer to have a sales rep. This situation could be indicated in the *CUSTOMER* table by setting such a customer's sales rep number to null. Technically, however, a null sales rep number would violate the restrictions that we have indicated for a foreign key. So if we were to use a null sales rep number, we would have to modify the definition of foreign keys to include the possibility of nulls. We would insist, though, that if the foreign key contained a value *other than null*, it would have to match the value of the primary key in some row in the other table. (In our example, for instance, a customer's sales rep number could be null, but if it were not, then it would have to be the number of an actual sales rep.) The general property we have just described is called referential integrity.

Definition: **Referential integrity** is the rule that if table A contains a foreign key that matches the primary key of table B then values of this foreign key either must match the value of the primary key for some row in table B or must be null.

The problems just mentioned are solved through the use of foreign keys. Indicating that the *SLSRNUMB* in the *CUSTOMER* table is a foreign key that must match the *SLSREP* table makes the relationship between customers and sales reps explicit. We do not need to look for common columns in several tables. Further, with foreign keys, matching columns that have different names no longer pose a problem. For example, it would not matter if the name of the foreign key in the *CUSTOMER* table happened to be *SLSRNUMB* while the primary key in the *SLSREP* table happened to be *SLSRNUMB*, the only thing that would matter is that this column was a foreign key that matched the *SLSREP* table. Finally, through referential integrity, it is possible for a customer not to have a sales rep number, but it is not possible for a customer to have *an invalid sales rep number*; that is, a customer's sales rep number *must* either be null or the number of a sales rep who is already in the database.

For other perspectives on integrity in the relational model, see [9] and [14].

4.6   CHANGING THE STRUCTURE OF A RELATIONAL DATABASE

One of the best things about relational DBMSs is the ease with which the database structure can be changed. New tables can be added and old ones can be removed. Columns can be added or deleted. The characteristics of columns can be changed. New indexes can be created and old ones can be dropped. Though the exact manner in which these changes are accomplished varies from one system to another, many systems allow all of these changes to be made quickly and easily. Since SQL is so widely used, we will use it as a vehicle to illustrate the manner in which these changes may be accomplished.

Alter

Changing the structure of a table in SQL is accomplished through the ALTER table command. Virtually every implementation of SQL allows new columns to be added to the end of an existing table. For example, let's suppose that we now wish to maintain a customer type for each customer in the *PREMIERE PRODUCTS* database. We can decide to call regular customers type R, distributors type D, and special customers type S. To implement this change, we need a new column in the customer table. This can be added as follows:

```
ALTER TABLE CUSTOMER
     ADD CUSTTYPE        CHAR(1)
```

At this point, the *CUSTOMER* table contains an extra column, *CUSTTYPE*.

For rows added from this point on, the value of *CUSTTYPE* will be assigned as the row is added. For existing rows, some value of *CUSTTYPE* must be assigned. The simplest approach (from the point of view of the DBMS, *not* the user) is to assign the value NULL as a *CUSTTYPE* on all existing rows. (This requires that *CUSTTYPE* accept null values, and some systems do require this. This means that any column added to a table definition *will* accept nulls; the user has no choice in the matter.) A more flexible approach, and one that is supported by some systems, is to allow the user to specify an initial value. In our example, if most customers were type R, we might set all of the customer types for existing customers to R and later change those customers of type D or type S to the appropriate value. To change the structure and set the value of *CUSTTYPE* to R for all existing records, we would type:

```
ALTER TABLE CUSTOMER
     ADD CUSTTYPE        CHAR(1)    INIT = 'R'
```

Some systems allow existing columns to be deleted. The syntax for deleting the *WREHSENM* column from the *PART* table would typically be something like this:

```
ALTER TABLE PART
     DELETE WREHSENM
```

Finally, some systems allow changes in the data types of given columns. A typical use of such a provision would be to increase the length of a character field that was found to be inadequate. Assuming that the *NAME* column in the *CUSTOMER* table needed to be increased to 30 characters, the ALTER statement would be something like this:

```
ALTER TABLE CUSTOMER
     CHANGE COLUMN NAME TO CHAR(30)
```

Drop

A table that is no longer needed can be deleted with the DROP command. If the *SLSREP* table were no longer needed in the *PREMIERE PRODUCTS* database, the command would be:

```
DROP TABLE SLSREP
```

The table would be erased, as would all indexes and views defined on the table. References to the table would be removed from the system catalog.

4.7 WHAT DOES IT TAKE TO BE RELATIONAL?

This chapter will conclude with a discussion of this interesting question. If you look at ads for both microcomputer and mainframe DBMSs, you will rarely find one that doesn't claim the DBMS is "relational." In order to make some sense out of all these claims, we need a yardstick to measure them with. In other words, we need to know what it really means for a system to be relational. For the answer to this question, let's first turn to the person who initially proposed the relational model, Dr. E. F. Codd.

In [7], Codd defines a relational system as one in which at least the following two properties hold:

1. Users perceive databases as collections of tables and are not aware of the presence of any additional structures.
2. The operations of SELECT, PROJECT, and JOIN from the relational algebra are supported. This support is independent of any predefined access paths; that is, it makes no difference which indexes do or do not exist. If, for example, a join can be performed only when indexes exist for the columns on which the join is to take place, the system should not be considered relational.

Codd's definition states that for a system to be relational, it must support the SELECT, PROJECT, and JOIN operations of the **relational algebra**, but the definition does *not* indicate that the system must use this terminology. SQL supports these three operations through the SELECT statement, which is considerably more powerful than the relational algebra SELECT indicated in Codd's definition. The following are examples of the manner in which SQL supports the SELECT, PROJECT, and JOIN operations.

SELECT (choose certain rows from a table):

```
SELECT *
    FROM CUSTOMER
    WHERE CREDLIM = 500
```

PROJECT (choose certain columns from a table):

```
SELECT CUSTNUMB, NAME, ADDRESS
    FROM CUSTOMER
```

JOIN (combine tables based on matching columns):

```
SELECT SLSRNUMB, SLSRNAME, SLSRADDR,
    TOTCOMM, COMMRATE, CUSTNUMB,
    NAME, ADDRESS, BALANCE, CREDLIM
    FROM SLSREP, CUSTOMER
    WHERE SLSREP.SLSRNUMB =
        CUSTOMER.SLSRNUMB
```

Let's turn now to C. J. Date, another individual who has been heavily involved in the development of the relational model. In [9], Date discusses a classification scheme for systems that support at least the relational structure, that is, systems in which the only structure that the user perceives is the table. Such systems fall into one of the following four categories:

1. **Tabular system.** In a **tabular system**, the only structure perceived by the user is the table, but the system does not support the SELECT, PROJECT, and JOIN operations in the unrestricted fashion indicated by Codd. (In this case, either the system does not support SELECT, PROJECT, and JOIN, or, if it does, the support

relies on predefined access paths.) The microcomputer file-management systems and some of the database management systems are in this category.

2. **Minimally relational.** Actually, the operations SELECT, PROJECT, and JOIN are not the only operations in the relational algebra. There are eight operations altogether. A **minimally relational** system is one that supports the tabular structure together with the SELECT, PROJECT, and JOIN operations but that does not support the complete set of operations from the relational algebra. The vast majority of microcomputer DBMSs fall into this category.

3. **Relationally complete.** Any system that supports the tabular structure and all the operations of the relational algebra (without requiring appropriate indexes) is said to be **relationally complete**. Many relational mainframe DBMSs and some microcomputer DBMSs fall into this category. In particular, any system that supports a *full* implementation of SQL is relationally complete.

4. **Fully relational.** A system that supports the tabular structure, all the operations of the relational algebra, and the two integrity rules (entity and referential integrity) described earlier in this section is said to be **fully relational**. This is the goal for which systems are (or should be) striving. Currently, the principal failing on the part of many systems is lack of support for referential integrity. Much progress is occurring in this area, and soon we shall see a number of fully relational systems.

SUMMARY

1. Views are used to give each user his or her own picture of the database.
   a. A view is defined in SQL through the use of a defining query.
   b. When a query is entered which references a view, it is merged with the defining query to produce the query that is actually executed.
   c. Retrieving data through a view presents no problem, but updating the database through a view is often prohibited.

2. Indexes are often used to facilitate retrieval. Indexes may be created on any column or combination of columns.

3. The catalog is a feature of many relational model DBMSs which stores information about the structure of a database. The system updates the catalog automatically. Users can retrieve data from the catalog in the same manner in which they retrieve data from the database.

4. There are two special integrity rules for relational databases:
   a. Entity integrity is the property that no column that is part of the primary key can accept null values.
   b. Referential integrity is the property that the value in any foreign key either must be null or must match an actual value of the primary key of another table.

5. Relational DBMSs provide facilities that allow users to easily change the structure of a database. Two examples of such facilities are as follows:
   a. ALTER allows columns to be added to a table, columns to be deleted, or characteristics of columns to be changed.
   b. DROP allows a table to be deleted from a database.

6. According to Codd, a DBMS cannot be considered to be relational unless the following two conditions pertain:
   a. Users perceive a database as simply a collection of tables.
   b. The DBMS supports at least the SELECT, PROJECT, and JOIN operations of the relational algebra.

7. According to Date, DBMSs in which users perceive databases as collections of tables can be classified as:
   a. Tabular, if data is viewed as tables and if the system does not support SELECT, PROJECT, and JOIN independently of any predefined access paths.
   b. Minimally relational, if SELECT, PROJECT, and JOIN are supported but the full set of operations in the relational algebra is not.
   c. Relationally complete, if the DBMS supports all operations of the relational algebra.
   d. Fully relational, if the DBMS supports all operations of the relational algebra and both integrity rules (entity and referential integrity).

KEY TERMS

catalog	minimally relational
defining query	null
entity integrity	optimizer
foreign key	referential integrity
fully relational	relationally complete
index	tabular
integrity rules	view

EXERCISES

1. What is a view? How is it defined? Does the data described in a view definition ever exist in that form? What happens when a user accesses a database through a view?
2. Define a view called *SMALLCUST*. It consists of the customer number, name, address, balance, and credit limit for all customers whose credit limit is $500 or less.
   a. Write the view definition for *SMALLCUST*.
   b. Write an SQL query to retrieve the number and name of all customers in *SMALLCUST* whose balance is over their credit limit.
   c. Convert the query from (b) to the query that will actually be executed.
3. Define a view called *CUSTORD*. It consists of the customer number, name, balance, order number, and order date for all orders currently on file.
   a. Write the view definition for *CUSTORD*.
   b. Write an SQL query to retrieve the customer number, name, order number, and order date for all orders in *CUSTORD* for customers whose balance is more than $100.
   c. Convert the query from (b) to the query that will actually be executed.
4. What are the advantages of using indexes? The disadvantages?
5. On relational mainframe DBMSs, who or what is responsible for the decision to use a particular index? What about on microcomputer DBMSs?
6. What is the catalog? What are three items about which the catalog maintains information?
7. Why is it a good idea for the DBMS to update the catalog automatically when a change is made in the database structure? Could users cause problems by updating the catalog themselves?
8. State the two integrity rules. Indicate the reasons for enforcing each rule.
9. How can the structure of a table be changed in SQL? What general types of changes are possible? Which commands are used to implement these changes?
10. List the two properties specified by Codd which a system must satisfy in order to be considered relational.
11. List the four categories of systems proposed by Date and describe the characteristics of systems in each category.

5 Database Design I: Normalization

OBJECTIVES

1. Present the idea of functional dependence.
2. Define the term primary key.
3. Define first normal form (1NF), second normal form (2NF), and third normal form (3NF).
4. Describe the problems associated with relations (tables) that are not in 1NF, 2NF, or 3NF, along with the mechanism for converting to all three.
5. Discuss the problems associated with incorrect conversions to 3NF.

5.1 INTRODUCTION

We have discussed the basic relational model, its structure, and the various ways of manipulating data within a relational database. In this chapter, we discuss the **normalization** process and its underlying concepts and features. Normalization enables us to analyze the design of a relational database to see whether it is bad; that is, normalization gives us a method for identifying the existence of potential problems, called **update anomalies**, in the design. The normalization process also supplies methods for correcting these problems.

The process involves various types of **normal forms**. **First normal form** (1NF), **second normal form** (2NF), and **third normal form** (3NF) are three of these types. It is these three that will be of the greatest use to us during database design. They form a progression in which a table that is in 1NF is better than a table that is not in 1NF; a table that is in 2NF is better yet; and so on. The goal of this process is to allow us to start with a table or collection of tables and produce a new collection of tables that is equivalent to the original collection (i.e., that represents the same information) but is free of problems. For practical purposes, this means that tables in the new collection will be in 3NF.

These normal forms were initially defined by Codd in 1972 (see [4]). Subsequently, it was discovered that the definition of third normal form was inadequate for certain situations. A revised and stronger definition was provided by Boyce and Codd in 1974 (see [5]). It is this more recent definition of third normal form (sometimes called **Boyce-Codd normal form**) that we shall examine later in this section.

We begin by discussing two crucial concepts that are fundamental to the understanding of the normalization process: functional dependence and keys. We then discuss first, second, and third normal forms. Finally, we look at the application of normalization to database design. We will then be ready to begin our study of the database design process in the next chapter.

Many of the examples in this chapter use data from the *PREMIERE PRODUCTS* example (see Figure 5.1).

Figure 5.1

PREMIERE PRODUCTS sample data

SLSREP

SLSRNUMB	SLSRNAME	SLSRADDR	TOTCOMM	COMMRATE
3	MARY JONES	123 MAIN,GRANT,MI	2150.00	.05
6	WILLIAM SMITH	102 RAYMOND,ADA,MI	4912.50	.07
12	SAM BROWN	419 HARPER,LANSING,MI	2150.00	.05

CUSTOMER

CUSTNUMB	CUSTNAME	ADDRESS	BALANCE	CREDLIM	SLSRNUMB
124	SALLY ADAMS	481 OAK,LANSING,MI	418.75	500	3
256	ANN SAMUELS	215 PETE,GRANT,MI	10.75	800	6
311	DON CHARLES	48 COLLEGE,IRA,MI	200.10	300	12
315	TOM DANIELS	914 CHERRY,KENT,MI	320.75	300	6
405	AL WILLIAMS	519 WATSON,GRANT,MI	201.75	800	12
412	SALLY ADAMS	16 ELM,LANSING,MI	908.75	1000	3
522	MARY NELSON	108 PINE,ADA,MI	49.50	800	12
567	JOE BAKER	808 RIDGE,HARPER,MI	201.20	300	6
587	JUDY ROBERTS	512 PINE,ADA,MI	57.75	500	6
622	DAN MARTIN	419 CHIP,GRANT,MI	575.50	500	3

ORDERS

ORDNUMB	ORDDTE	CUSTNUMB
12489	90294	124
12491	90294	311
12494	90494	315
12495	90494	256
12498	90594	522
12500	90594	124
12504	90594	522

ORDLNE

ORDNUMB	PARTNUMB	NUMBORD	QUOTPRCE
12489	AX12	11	14.95
12491	BT04	1	402.99
12491	BZ66	1	311.95
12491	CB03	4	175.00
12495	CX11	2	57.95
12498	AZ52	2	22.95
12498	BA74	4	4.95
12500	BT04	1	402.99
12504	CZ81	2	108.99

PART

PARTNUMB	PARTDESC	UNONHAND	ITEMCLSS	WREHSENM	UNITPRCE
AX12	IRON	104	HW	3	17.95
AZ52	SKATES	20	SG	2	24.95
BA74	BASEBALL	40	SG	1	4.95
BH22	TOASTER	95	HW	3	34.95
BT04	STOVE	11	AP	2	402.99
BZ66	WASHER	52	AP	3	311.95
CA14	SKILLET	2	HW	3	19.95
CB03	BIKE	44	SG	1	187.50
CX11	MIXER	112	HW	3	57.95
CZ81	WEIGHTS	208	SG	2	108.99

5.2 FUNCTIONAL DEPENDENCE

The concept of functional dependence is crucial to the material in the rest of this chapter. Functional dependence is a fancy name for what is basically a simple idea. To illustrate it, suppose that the *SLSREP* table for *PREMIERE PRODUCTS* is as shown in Figure 5.2.

The only difference between this *SLSREP* table and the one we have been looking at previously is the addition of an extra column, *PAYCLASS* (pay class).

Figure 5.2

SLSREP table with additional column, *PAYCLASS*

SLSREP

SLSRNUMB	SLSRNAME	SLSRADDR	TOTCOMM	PAYCLASS	COMMRATE
3	MARY JONES	123 MAIN,GRANT,MI	2150.00	1	.05
6	WILLIAM SMITH	102 RAYMOND,ADA,MI	4912.50	2	.07
12	SAM BROWN	419 HARPER,LANSING,MI	2150.00	1	.05

Let's suppose further that one of the policies at Premiere Products is that all sales reps in any given pay class get the same commission rate. If you were asked to describe this policy in another way, you might say something like, "A sales rep's pay class *determines* his or her commission rate." Or you might say, "A sales rep's commission rate *depends on* his or her pay class." If you said either of these things, you would be using the word *determines* or the words *depends on* in exactly the fashion that we will be using them. If we wanted to be formal, we would precede either expression with the word *functionally*. Thus we might say, "A sales rep's pay class *functionally determines* his or her commission rate," or "A sales rep's commission rate *functionally depends on* his or her pay class." The formal definition of functional dependence is as follows:

> *Definition:* An attribute, B, is **functionally dependent** on another attribute, A (or possibly a collection of attributes), if a value for A determines a single value for B at any one time.

We can think of this as follows. If we are given a value for A, do we know that we will be able to find a single value for B? If so, B is functionally dependent on A (often written as A $-->$ B). If B is functionally dependent on A, we also say that A **functionally determines** B.

For example, in the *CUSTOMER* table, is the *CUSTNAME* functionally dependent on *CUSTNUMB*? The answer is yes. If we are given customer number 124, for example, we would find a *single* name, Sally Adams, associated with it.

In the same *CUSTOMER* table, is *ADDRESS* functionally dependent on *CUSTNAME*? Here the answer is no since, given the name Sally Adams, we would not be able to find a single address.

In the *ORDLNE* table, is the *NUMBORD* functionally dependent on *ORDNUMB*? No. *ORDNUMB* does not give enough information. Is it functionally dependent on *PARTNUMB*? No. Again, not enough information is given. In reality, *NUMBORD* is functionally dependent on the **concatenation** (combination) of *ORDNUMB* and *PARTNUMB*.

At this point, a question naturally arises: How do we determine functional dependencies? Can we determine them by looking at sample data? The answer is no.

Consider Figure 5.3, in which customer names happen to be unique. It is very tempting to say that *CUSTNAME* functionally determines *ADDRESS* (or equivalently that *ADDRESS* is functionally dependent on *CUSTNAME*). After all, given the name of a customer, we can find the single address. But what happens when customer 412, whose name also happens to be Sally Adams, is added to the database? We then have the situation exhibited in Figure 5.4.

Figure 5.3

CUSTOMER table

CUSTOMER

CUSTNUMB	CUSTNAME	ADDRESS	BALANCE	CREDLIM	SLSRNUMB
124	SALLY ADAMS	481 OAK,LANSING,MI	418.75	500	3
256	ANN SAMUELS	215 PETE,GRANT,MI	10.75	800	6
311	DON CHARLES	48 COLLEGE,IRA,MI	200.10	300	12
315	TOM DANIELS	914 CHERRY,KENT,MI	320.75	300	6
405	AL WILLIAMS	519 WATSON,GRANT,MI	201.75	800	12
522	MARY NELSON	108 PINE,ADA,MI	49.50	800	12
567	JOE BAKER	808 RIDGE,HARPER,MI	201.20	300	6
587	JUDY ROBERTS	512 PINE,ADA,MI	57.75	500	6
622	DAN MARTIN	419 CHIP,GRANT,MI	575.50	500	3

Figure 5.4

CUSTOMER table
with second Sally
Adams

CUSTOMER

CUSTNUMB	CUSTNAME	ADDRESS	BALANCE	CREDLIM	SLSRNUMB
124	SALLY ADAMS	481 OAK,LANSING,MI	418.75	500	3
256	ANN SAMUELS	215 PETE,GRANT,MI	10.75	800	6
311	DON CHARLES	48 COLLEGE,IRA,MI	200.10	300	12
315	TOM DANIELS	914 CHERRY,KENT,MI	320.75	300	6
405	AL WILLIAMS	519 WATSON,GRANT,MI	201.75	800	12
412	SALLY ADAMS	16 ELM,LANSING,MI	908.75	1000	3
522	MARY NELSON	108 PINE,ADA,MI	49.50	800	12
567	JOE BAKER	808 RIDGE,HARPER,MI	201.20	300	6
587	JUDY ROBERTS	512 PINE,ADA,MI	57.75	500	6
622	DAN MARTIN	419 CHIP,GRANT,MI	575.50	500	3

If the name we are given is Sally Adams, we can no longer find a single address. Thus we were misled by our original sample data. The only way to really determine the functional dependencies that exist is to examine the user's policies.

5.3   KEYS

A second underlying concept of the normalization process is that of the primary key. It builds on functional dependence, and it completes the background required for an understanding of the normal forms.

> *Definition:* Attribute A (or a collection of attributes) is the **primary key** for a relation (table), R, if
>
> 1. *All* attributes in R are functionally dependent on A.
> 2. No subcollection of the attributes in A (assuming A is a collection of attributes and not just a single attribute) also has property 1.

For example, is *CUSTNAME* the primary key for the *CUSTOMER* table? No, since the other attributes are not functionally dependent on name. (Note that the answer would be different in an organization that had a policy enforcing uniqueness of customer names.)

Is *CUSTNUMB* the primary key for the *CUSTOMER* table? Yes, since all attributes in the *CUSTOMER* table are functionally dependent on *CUSTNUMB*.

Is *ORDNUMB* the primary key for the *ORDLNE* table? No, since it does not uniquely determine *NUMBORD* or *QUOTPRCE*.

Is the combination of the *ORDNUMB* and the *PARTNUMB* the primary key for the *ORDLNE* table? Yes, since all attributes can be determined by this combination, and nothing less will do.

Is the combination of the *PARTNUMB* and the *PARTDESC* the primary key for the *PART* table? No. Though it is true that all attributes of the *PART* table can be determined by this combination, something less, namely, the *PARTNUMB* alone, also has this property.

Occasionally (but not often) there might be more than one possibility for the primary key. For example, in an *EMPLOYEE* table either the *EMPNUMB* or the *SSNUMB* (social security number) could serve as the key. In this case one of these is designated as the primary key. The other is referred to as a **candidate key**. A candidate key is a collection of attributes that has the same properties presented in the definition of the primary key. (Technically, the definition given for primary key really defines candidate key. From all of the candidate keys one is chosen to be the primary key. The candidate keys that are not chosen to be the primary key are often referred to as **alternate keys**.)

Note: The primary key is frequently called simply the *key* in other studies on database management and the relational model. We will continue to use the term primary key in order to clearly distinguish among the several different concepts of a key that we will encounter.

5.4 FIRST NORMAL FORM

A relation (table) that contains a repeating group is called an **unnormalized relation**. (Technically, it is not a relation at all.) Removal of repeating groups is the starting point in our quest for relations that are as free of problems as possible. Relations without repeating groups are said to be in first normal form.

Definition: A relation (table) is in **first normal form** (1NF) if it does not contain repeating groups.

As an example, consider the following *ORDERS* table, in which there is a repeating group consisting of *PARTNUMB* and *NUMBORD*. As the example shows, there is one row per order with *PARTNUMB, NUMBORD* repeated as many times as is necessary.

ORDERS (<u>ORDNUMB</u>, ORDDTE, PARTNUMB, NUMBORD)

(This notation indicates a table called *ORDERS*, consisting of a primary key, *ORDNUMB*, an attribute, *ORDDTE*, and a repeating group containing two attributes, *PARTNUMB* and *NUMBORD*.) Figure 5.5 shows a sample of this table.

Figure 5.5

Sample unnormalized table

ORDERS

ORDNUMB	ORDDTE	PARTNUMB	NUMBORD
12489	90294	AX12	11
12491	90294	BT04	1
		BZ66	1
12494	90494	CB03	4
12495	90494	CX11	2
12498	90594	AZ52	2
		BA74	4
12500	90594	BT04	1
12504	90594	CZ81	2

To convert this table to 1NF, the repeating group is removed, giving the following:

ORDERS(<u>ORDNUMB</u>, ORDDTE, <u>PARTNUMB</u>, NUMBORD)

The corresponding example of the new table is shown in Figure 5.6.

Figure 5.6

Result of
normalization
(conversion to 1NF)

ORDERS

ORDNUMB	ORDDTE	PARTNUMB	NUMBORD
12489	90294	AX12	11
12491	90294	BT04	1
12491	90294	BZ66	1
12494	90494	CB03	4
12495	90494	CX11	2
12498	90594	AZ52	2
12498	90594	BA74	4
12500	90594	BT04	1
12504	90594	CZ81	2

Note that the second row of the unnormalized table indicates that part BZ66 and part BT04 are both present for order 12491. In the normalized table, this information is represented by *two* rows, the second and third. The primary key to the unnormalized *ORDERS* table was the *ORDNUMB* alone. The primary key to the normalized table is now the combination of *ORDNUMB* and *PARTNUMB*. In general it will be true that the primary key will expand in converting a non-1NF table to 1NF. It will typically include the original primary key concatenated with the key to the repeating group; i.e., the attribute that distinguishes one occurrence of the repeating group from another within a given row in the table. In this case, *PARTNUMB* is the key to the repeating group and thus becomes part of the primary key of the 1NF table.

5.5 *SECOND NORMAL FORM*

Even though the following table is in 1NF, problems exist within the table that will cause us to want to restructure it. Consider the table:

ORDERS(<u>ORDNUMB</u>, ORDDTE, <u>PARTNUMB</u>, PARTDESC, NUMBORD, QUOTPRCE)

with the functional dependencies:

ORDNUMB --> ORDDTE
PARTNUMB --> PARTDESC
ORDNUMB, PARTNUMB --> NUMBORD, QUOTPRCE

Thus *ORDNUMB* determines *ORDDTE*, *PARTNUMB* determines *PARTDESC*, and the concatenation of *ORDNUMB* and *PARTNUMB* determines *NUMBORD* and *QUOTPRCE*. Consider the sample of this table shown in Figure 5.7.

Figure 5.7

Sample *ORDERS*
table

ORDERS

ORDNUMB	ORDDTE	PARTNUMB	PARTDESC	NUMBORD	QUOTPRCE
12489	90294	AX12	IRON	11	14.95
12491	90294	BT04	STOVE	1	402.99
12491	90294	BZ66	WASHER	1	311.95
12494	90494	CB03	BIKE	4	175.00
12495	90494	CX11	MIXER	2	57.95
12498	90594	AZ52	SKATES	2	22.95
12498	90594	BA74	BASEBALL	4	4.95
12500	90594	BT04	STOVE	1	402.99
12504	90594	CZ81	WEIGHTS	2	108.99

As you can see in the example, the description of a specific part, BT04 for example, occurs several times in the table. This redundancy causes several problems. It is certainly wasteful of space, but that in itself is not nearly as serious as some of the other problems. These other problems are called **update anomalies** and they fall into four categories:

1. **Update.** A change to the description of part BT04 requires not one change but several—we have to change each row in which BT04 appears. This certainly makes the update process much more cumbersome; it is more complicated logically and takes more time to update.

2. **Inconsistent data.** There is nothing about the design that would prohibit part BT04 from having two different descriptions in the database. In fact, if it occurs in twenty rows, it could conceivably have twenty *different* descriptions in the database!!!

3. **Additions.** We have a real problem when we try to add a new part and its description to the database. Since the primary key for the table consists of both *ORDNUMB* and *PARTNUMB*, we need values for both of these in order to add a new row. If we have a part to add but there are as yet no orders for it, what do we use for an *ORDNUMB*? Our only solution would be to make up a dummy order number and then replace it with a real *ORDNUMB* once an order for this part had actually been received. Certainly this is not an acceptable solution!

4. **Deletions.** In the example above, if we delete order 12489 from the database, we also *lose* the fact that part AX12 is called IRON.

The problems just described occur because we have an attribute, *PARTDESC*, that is dependent on only a portion of the primary key, *PARTNUMB*, and *not* on the complete primary key. This leads to the definition of second normal form. Second normal form represents an improvement over first normal form since it eliminates these update anomalies in these situations. First, we need to define nonkey attribute.

Definition: An attribute is a **nonkey attribute** if it is not a part of the primary key.

We can now provide a definition for second normal form.

Definition: A relation (table) is in **second normal form** (2NF) if it is in first normal form and no nonkey attribute is dependent on only a portion of the primary key.

For another perspective on 2NF, consider Figure 5.8. This type of diagram, sometimes called a **dependency diagram**, indicates all of the functional dependencies present in the *ORDERS* table through arrows. The arrows above the boxes indicate the normal dependencies that should be present; i.e., the primary key functionally determines all other attributes. (In this case, the concatenation of *ORDNUMB* and *PARTNUMB* determines all other attributes.) It is the arrows below the boxes that prevent the table from being in 2NF. These arrows represent what is often termed **partial dependencies**, which are dependencies on something less than the key. In fact, an alternative definition for 2NF is that a table is in 2NF if it is in 1NF but contains no partial dependencies.

Figure 5.8

Dependencies in
ORDERS table
(blue arrows
represent partial
dependencies)

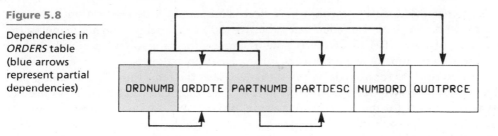

Either way we view 2NF, we can now name the fundamental problem with the *ORDERS* table: it is *not* in 2NF. While it may be pleasing to have a name for the problem, what we really need, of course, is a method to *correct* it. Such a method follows.

First, for each subset of the set of attributes that make up the primary key, begin a table with this subset as its primary key. For the *ORDERS* table, this would give:

    (ORDNUMB,
    (PARTNUMB,
    (ORDNUMB, PARTNUMB,

Next, place each of the other attributes with the appropriate primary key; that is, place each one with the minimal collection on which it depends. For the *ORDERS* table this would yield:

    (ORDNUMB, ORDDTE)
    (PARTNUMB, PARTDESC)
    (ORDNUMB, PARTNUMB, NUMBORD, QUOTPRCE)

Each of these tables can now be given a name that is descriptive of the meaning of the table, such as *ORDERS*, *PART*, or *ORDLNE*, for example. Figure 5.9 shows samples of the tables involved.

Figure 5.9

Conversion to 2NF

ORDERS

ORDNUMB	ORDDTE	PARTNUMB	PARTDESC	NUMBORD	QUOTPRCE
12489	90294	AX12	IRON	11	14.95
12491	90294	BT04	STOVE	1	402.99
12491	90294	BZ66	WASHER	1	311.95
12494	90494	CB03	BIKE	4	175.00
12495	90494	CX11	MIXER	2	57.95
12498	90594	AZ52	SKATES	2	22.95
12498	90594	BA74	BASEBALL	4	4.95
12500	90594	BT04	STOVE	1	402.99
12504	90594	CZ81	WEIGHTS	2	108.99

is replaced by

ORDERS

ORDNUMB	ORDDTE
12489	90294
12491	90294
12494	90494
12495	90494
12498	90594
12500	90594
12504	90594

PART

PARTNUMB	PARTDESC
AX12	IRON
AZ52	SKATES
BA74	BASEBALL
BH22	TOASTER
BT04	STOVE
BZ66	WASHER
CA14	SKILLET
CB03	BIKE
CX11	MIXER
CZ81	WEIGHTS

ORDLNE

ORDNUMB	PARTNUMB	NUMBORD	QUOTPRCE
12489	AX12	11	14.95
12491	BT04	1	402.99
12491	BZ66	1	311.95
12494	CB03	4	175.00
12495	CX11	2	57.95
12498	AZ52	2	22.95
12498	BA74	4	4.95
12500	BT04	1	402.99
12504	CZ81	2	108.99

Note that the update anomalies have been eliminated. A description appears only once, so we do not have the redundancy that we did in the earlier design. Changing the description of part BT04 to OVEN is now a simple process involving a single change. Since the description for a part occurs in one single place, it is not possible to have multiple descriptions for a single part in the database at the same time. To add a new part and its description, we create a new row in the *PART* table and thus there is no need to have an order exist for that part. Also, deleting order 12489 does not cause part number AX12 to be deleted from the *PART* table, and thus we still have its description (IRON) in the database. Finally, we have not lost any information in the process. The data in the original design can be reconstructed from the data in the new design.

5.6 THIRD NORMAL FORM

Problems can still exist with tables that are in 2NF. Consider the following *CUSTOMER* table:

```
CUSTOMER(CUSTNUMB, CUSTNAME, ADDRESS, SLSRNUMB, SLSRNAME)
```

with the functional dependencies:

```
CUSTNUMB --> CUSTNAME, ADDRESS, SLSRNUMB, SLSRNAME
SLSRNUMB --> SLSRNAME
```

(*CUSTNUMB* determines all of the other attributes. In addition *SLSRNUMB* determines *SLSRNAME*.)

If the primary key of a table is a single column, the table is automatically in second normal form. (If the table were not in 2NF, some column would be dependent on only a *portion* of the primary key, which is impossible when the primary key is just one column.) Thus, the *CUSTOMER* table is in second normal form.

As the sample of this table, shown in Figure 5.10, demonstrates, this table possesses problems similar to those encountered earlier, even though it is in second normal form. In this case it is the name of a sales rep that can occur many times in the table; see sales rep 12 (Sam Brown), for example. This redundancy results in the same exact set of problems that was described in the previous *ORDERS* table.

Figure 5.10

Sample *CUSTOMER* table

CUSTOMER

CUSTNUMB	CUSTNAME	ADDRESS	SLSRNUMB	SLSRNAME
124	SALLY ADAMS	481 OAK,LANSING,MI	3	MARY JONES
256	ANN SAMUELS	215 PETE,GRANT,MI	6	WILLIAM SMITH
311	DON CHARLES	48 COLLEGE,IRA,MI	12	SAM BROWN
315	TOM DANIELS	914 CHERRY,KENT,MI	6	WILLIAM SMITH
405	AL WILLIAMS	519 WATSON,GRANT,MI	12	SAM BROWN
412	SALLY ADAMS	16 ELM,LANSING,MI	3	MARY JONES
522	MARY NELSON	108 PINE,ADA,MI	12	SAM BROWN
567	JOE BAKER	808 RIDGE,HARPER,MI	6	WILLIAM SMITH
587	JUDY ROBERTS	512 PINE,ADA,MI	6	WILLIAM SMITH
622	DAN MARTIN	419 CHIP,GRANT,MI	3	MARY JONES

In addition to the problem of wasted space, we have similar update anomalies, as follows:

1. **Update.** A change to the name of a sales rep requires not one change but several. Again the update process becomes very cumbersome.
2. **Inconsistent data.** There is nothing about the design that would prohibit a sales rep from having two different names in the database. In fact, if the same sales rep

represents twenty different customers (and thus would be found on twenty different rows), he or she could have twenty different names in the database.

3. **Additions.** In order to add sales rep 47, whose name is Mary Daniels, to the database, we must have at least one customer whom she represents. If she has not yet been assigned any customers, then either we cannot record the fact that her name is Mary Daniels or we have to create a fictitious customer for her to represent. Again, this is not a very desirable solution to the problem.

4. **Deletions.** If we were to delete all of the customers of sales rep 6 from the database, then we would also lose the name of sales rep 6.

These update anomalies are due to the fact that *SLSRNUMB* determines *SLSRNAME* but *SLSRNUMB* is not the primary key. As a result, the same *SLSRNUMB* and consequently the same *SLSRNAME* can appear on many different rows.

We've seen that 2NF is an improvement over 1NF, but in order to eliminate 2NF problems, we need an even better strategy for creating tables in our database. Third normal form gives us that strategy. Before we look at third normal form, however, we need to become familiar with the special name that is given to any column that determines another column (like *SLSRNUMB* in the *CUSTOMER* table).

Definition: Any attribute (or collection of attributes) that determines another attribute is called a **determinant**.

Certainly the **primary key** in a table will be a determinant. In fact, by definition, any **candidate key** will be a determinant. (Remember that a candidate key is an attribute or collection of attributes which could have functioned as the primary key.) In this case, *SLSRNUMB* is a determinant but it is certainly not a candidate key, and that is the problem.

Definition: A relation (table) is in **third normal form** (3NF) if it is in second normal form and if the only determinants it contains are candidate keys.

Again, for an additional perspective, we will consider a dependency diagram, as shown in Figure 5.11. As before, the arrows above the boxes represent the normal dependencies of all attributes on the primary key. It is the arrow below the boxes that causes the problem. The presence of this arrow makes *SLSRNUMB* a determinant. If there were arrows from *SLSRNUMB* to all of the attributes, *SLSRNUMB* would be a candidate key and we would not have a problem. The absence of these arrows indicates that this table possesses a determinant that is not a candidate key. Thus, the table is not in 3NF.

Figure 5.11

Dependencies in *CUSTOMER* table (*SLSRNUMB* is a determinant since it functionally determines *SLSRNAME*)

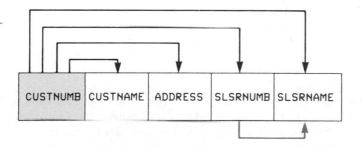

We have now named the problem with the *CUSTOMER* table: it is not in 3NF. What we need is a scheme to correct the deficiency in the *CUSTOMER* table and in all tables having similar deficiencies. Such a method follows.

First, for each determinant that is not a candidate key, remove from the table the attributes that depend on this determinant. Next, create a new table containing all the attributes from the original table that depend on this determinant. Finally, make the determinant the primary key of this new table.

In the *CUSTOMER* table, for example, *SLSRNAME* is removed since it depends on the determinant *SLSRNUMB*, which is not a candidate key. A new table is formed, consisting of *SLSRNUMB* as the primary key and *SLSRNAME*. Specifically

CUSTOMER(<u>CUSTNUMB</u>, CUSTNAME, ADDRESS, SLSRNUMB, SLSRNAME)

is replaced by

CUSTOMER(<u>CUSTNUMB</u>, CUSTNAME, ADDRESS, SLSRNUMB)

and

SLSREP(<u>SLSRNUMB</u>, SLSRNAME)

Figure 5.12 shows samples of the tables involved.

Figure 5.12

Conversion to 3NF

CUSTOMER

CUSTNUMB	CUSTNAME	ADDRESS	SLSRNUMB	SLSRNAME
124	SALLY ADAMS	481 OAK,LANSING,MI	3	MARY JONES
256	ANN SAMUELS	215 PETE,GRANT,MI	6	WILLIAM SMITH
311	DON CHARLES	48 COLLEGE,IRA,MI	12	SAM BROWN
315	TOM DANIELS	914 CHERRY,KENT,MI	6	WILLIAM SMITH
405	AL WILLIAMS	519 WATSON,GRANT,MI	12	SAM BROWN
412	SALLY ADAMS	16 ELM,LANSING,MI	3	MARY JONES
522	MARY NELSON	108 PINE,ADA,MI	12	SAM BROWN
567	JOE BAKER	808 RIDGE,HARPER,MI	6	WILLIAM SMITH
587	JUDY ROBERTS	512 PINE,ADA,MI	6	WILLIAM SMITH
622	DAN MARTIN	419 CHIP,GRANT,MI	3	MARY JONES

is replaced by

CUSTOMER

CUSTNUMB	CUSTNAME	ADDRESS	SLSRNUMB
124	SALLY ADAMS	481 OAK,LANSING,MI	3
256	ANN SAMUELS	215 PETE,GRANT,MI	6
311	DON CHARLES	48 COLLEGE,IRA,MI	12
315	TOM DANIELS	914 CHERRY,KENT,MI	6
405	AL WILLIAMS	519 WATSON,GRANT,MI	12
412	SALLY ADAMS	16 ELM,LANSING,MI	3
522	MARY NELSON	108 PINE,ADA,MI	12
567	JOE BAKER	808 RIDGE,HARPER,MI	6
587	JUDY ROBERTS	512 PINE,ADA,MI	6
622	DAN MARTIN	419 CHIP,GRANT,MI	3

SLSREP

SLSRNUMB	SLSRNAME
3	MARY JONES
6	WILLIAM SMITH
12	SAM BROWN

Have we now corrected all previously identified problems? A sales rep's name appears only once, thus avoiding redundancy and making the process of changing a sales rep's name a very simple one. It is not possible with this design for the same sales rep to have two different names in the database. To add a new sales rep to the database, we add a row in the *SLSREP* table so that it is not necessary to have a customer whom the sales rep represents. Finally, deleting all of the customers of a given sales rep will not remove the sales rep's record from the *SLSREP* table, so we do retain the sales rep's name; all of the data in the original table can be reconstructed from the data in the new collection of tables. All previously mentioned problems have indeed been solved.

5.7 INCORRECT DECOMPOSITIONS

It is important to note that the decomposition of a table into two or more 3NF tables *must* be accomplished by the method indicated even though there are other possibilities that might seem at first glance to be legitimate. Let us examine two other decompositions of the *CUSTOMER* table into 3NF tables in order to understand the difficulties they pose.

What if, in the decomposition process,

CUSTOMER(<u>CUSTNUMB</u>, CUSTNAME, ADDRESS, SLSRNUMB, SLSRNAME)

is replaced by

CUSTOMER(<u>CUSTNUMB</u>, CUSTNAME, ADDRESS, SLSRNUMB)

and

SLSREP(<u>CUSTNUMB</u>, SLSRNAME)

Samples of these tables are shown in Figure 5.13.

Figure 5.13

Incorrect decomposition

CUSTOMER

CUSTNUMB	CUSTNAME	ADDRESS	SLSRNUMB	SLSRNAME
124	SALLY ADAMS	481 OAK,LANSING,MI	3	MARY JONES
256	ANN SAMUELS	215 PETE,GRANT,MI	6	WILLIAM SMITH
311	DON CHARLES	48 COLLEGE,IRA,MI	12	SAM BROWN
315	TOM DANIELS	914 CHERRY,KENT,MI	6	WILLIAM SMITH
405	AL WILLIAMS	519 WATSON,GRANT,MI	12	SAM BROWN
412	SALLY ADAMS	16 ELM,LANSING,MI	3	MARY JONES
522	MARY NELSON	108 PINE,ADA,MI	12	SAM BROWN
567	JOE BAKER	808 RIDGE,HARPER,MI	6	WILLIAM SMITH
587	JUDY ROBERTS	512 PINE,ADA,MI	6	WILLIAM SMITH
622	DAN MARTIN	419 CHIP,GRANT,MI	3	MARY JONES

is replaced by

CUSTOMER

CUSTNUMB	CUSTNAME	ADDRESS	SLSRNUMB
124	SALLY ADAMS	481 OAK,LANSING,MI	3
256	ANN SAMUELS	215 PETE,GRANT,MI	6
311	DON CHARLES	48 COLLEGE,IRA,MI	12
315	TOM DANIELS	914 CHERRY,KENT,MI	6
405	AL WILLIAMS	519 WATSON,GRANT,MI	12
412	SALLY ADAMS	16 ELM,LANSING,MI	3
522	MARY NELSON	108 PINE,ADA,MI	12
567	JOE BAKER	808 RIDGE,HARPER,MI	6
587	JUDY ROBERTS	512 PINE,ADA,MI	6
622	DAN MARTIN	419 CHIP,GRANT,MI	3

SLSREP

CUSTNUMB	SLSRNAME
124	MARY JONES
256	WILLIAM SMITH
311	SAM BROWN
315	WILLIAM SMITH
405	SAM BROWN
412	MARY JONES
522	SAM BROWN
567	WILLIAM SMITH
587	WILLIAM SMITH
622	MARY JONES

Both of the new tables are in 3NF. In addition, by joining these two tables together on *CUSTNUMB* we can reconstruct the original *CUSTOMER* table. The result, however, still suffers from some of the same kinds of problems that the original *CUSTOMER* table did. Consider, for example, the redundancy in the storage of sales reps' names, the problem encountered in changing the name of sales rep 12, and the difficulty of adding a new sales rep for whom there are as yet no customers. In addition, since the sales rep number is in one table and the sales rep name is in another, we have actually *split a functional dependence across two different tables*. Thus, this decomposition, while it may appear to be valid, is definitely not a desirable way to create 3NF tables.

There is another decomposition that we might choose, and that is to replace

CUSTOMER(<u>CUSTNUMB</u>, CUSTNAME, ADDRESS, SLSRNUMB, SLSRNAME)

by

CUSTOMER(<u>CUSTNUMB</u>, CUSTNAME, ADDRESS, SLSRNAME)

and

SLSREP(<u>SLSRNUMB</u>, SLSRNAME)

Samples of these tables are shown in Figure 5.14.

Figure 5.14

Second incorrect
decomposition

CUSTOMER

CUSTNUMB	CUSTNAME	ADDRESS	SLSRNUMB	SLSRNAME
124	SALLY ADAMS	481 OAK,LANSING,MI	3	MARY JONES
256	ANN SAMUELS	215 PETE,GRANT,MI	6	WILLIAM SMITH
311	DON CHARLES	48 COLLEGE,IRA,MI	12	SAM BROWN
315	TOM DANIELS	914 CHERRY,KENT,MI	6	WILLIAM SMITH
405	AL WILLIAMS	519 WATSON,GRANT,MI	12	SAM BROWN
412	SALLY ADAMS	16 ELM,LANSING,MI	3	MARY JONES
522	MARY NELSON	108 PINE,ADA,MI	12	SAM BROWN
567	JOE BAKER	808 RIDGE,HARPER,MI	6	WILLIAM SMITH
587	JUDY ROBERTS	512 PINE,ADA,MI	6	WILLIAM SMITH
622	DAN MARTIN	419 CHIP,GRANT,MI	3	MARY JONES

is replaced by

CUSTOMER

CUSTNUMB	CUSTNAME	ADDRESS	SLSRNAME
124	SALLY ADAMS	481 OAK,LANSING,MI	MARY JONES
256	ANN SAMUELS	215 PETE,GRANT,MI	WILLIAM SMITH
311	DON CHARLES	48 COLLEGE,IRA,MI	SAM BROWN
315	TOM DANIELS	914 CHERRY,KENT,MI	WILLIAM SMITH
405	AL WILLIAMS	519 WATSON,GRANT,MI	SAM BROWN
412	SALLY ADAMS	16 ELM,LANSING,MI	MARY JONES
522	MARY NELSON	108 PINE,ADA,MI	SAM BROWN
567	JOE BAKER	808 RIDGE,HARPER,MI	WILLIAM SMITH
587	JUDY ROBERTS	512 PINE,ADA,MI	WILLIAM SMITH
622	DAN MARTIN	419 CHIP,GRANT,MI	MARY JONES

SLSREP

SLSRNUMB	SLSRNAME
3	MARY JONES
6	WILLIAM SMITH
12	SAM BROWN

This seems to be a possibility. Not only are both tables in 3NF, but joining them together based on *SLSRNAME* seems to reconstruct the data in the original table. Or does it? Suppose that the name of sales rep 6 is also Mary Jones. In that case, when we join the two new tables together, we will get a row in which customer 124 (Sally Adams) is associated with sales rep 3 (Mary Jones) and *another* row in which customer 124 is associated with sales rep 6 (William Smith). Since we obviously want decompositions that preserve the original information, this scheme is not appropriate.

Question:

Using the types of entities found in a college environment (faculty, students, departments, courses, etc.), create an example of a table that is in 1NF but not in 2NF and an example of a table that is in 2NF but not in 3NF. In each case justify the answers and show how to convert to the higher forms.

Answer: There are many possible solutions. If your solution differs from the one we will look at, this does not mean that it is an unsatisfactory solution.

To create a 1NF table that is not in 2NF, we need a table that (a) has no repeating groups and (b) has at least one attribute that is dependent on only a portion of the primary key. For an attribute to be dependent on a portion of the primary key, the key must contain at least two attributes. Following is a picture of what we need:

(<u> 1 </u>, <u> 2 </u>,   3   ,   4   )

This table contains four attributes, numbered 1, 2, 3, and 4, in which attributes 1 and 2 functionally determine both attributes 3 and 4. In addition, neither attribute 1 nor attribute 2 can determine *all* other attributes, otherwise the key would contain only this one attribute. Finally, we want part of the key, say, attribute 2, to determine another attribute, say, attribute 4. Now that we have the pattern we need, we would like to find attributes from within the college environment to fit it. One example would be:

(<u>STUNUMB</u>, <u>CRSENUMB</u>, GRADE, CRSEDESC)

In this example, the concatenation of *STUNUMB* (student number) and *CRSENUMB* (course number) determines both *GRADE* and *CRSEDESC* (course description). Both of these are required to determine *GRADE*, and thus the primary key consists of their concatenation (nothing less will do). The *CRSEDESC*, however, is only dependent on the *CRSENUMB*. This violates second normal form. To convert this table to 2NF we would replace it by the two tables

(<u>STUNUMB</u>, <u>CRSENUMB</u>, GRADE)

and

(<u>CRSENUMB</u>, CRSEDESC)

We would of course now give these tables appropriate names.

To create a table that is in 2NF but not in 3NF, we need a 2NF table in which there is a determinant that is *not* a candidate key. If we choose a table that has a single attribute as the primary key, it is automatically in 2NF, so the real problem is the determinant. We need a table like the following:

(<u>   1   </u>,   2   ,   3   )

This table contains three attributes, numbered 1, 2, and 3, in which attribute 1 determines each of the others and is thus the primary key. If, in addition, attribute 2 determines attribute 3, it is a determinant. If it does not also determine attribute 1, then it is not a candidate key. One example that fits this pattern would be:

(<u>STUNUMB</u>, ADVNUMB, ADVNAME)

Here the student number (*STUNUMB*) determines both the student's advisor's number (*ADVNUMB*) and the advisor's name (*ADVNAME*). *ADVNUMB* determines *ADVNAME* but *ADVNUMB* does not determine *STUNUMB*, since one advisor can have many advisees. This table is in 2NF but not in 3NF. To convert it to 3NF, we replace it by

(<u>STUNUMB</u>, ADVNUMB)

and

(<u>ADVNUMB</u>, ADVNAME)

Question:

Convert the following table to 3NF:

```
STUDENT(STUNUMB, STUNAME, NUMBCRED,
ADVNUMB, ADVNAME,
              CRSENUMB, CRSEDESC, GRADE)
```

In this table, *STUNUMB* determines *STUNAME, NUMBCRED, ADVNUMB,* and *ADVNAME. ADVNUMB* determines *ADVNAME. CRSENUMB* determines *CRSEDESC.* The combination of a *STUNUMB* and a *CRSENUMB* determines a *GRADE.*

Answer:

Step 1. Remove the repeating group to convert to 1NF. This yields:

```
STUDENT(STUNUMB, STUNAME, NUMBCRED,
          ADVNUMB, ADVNAME,
          CRSENUMB,
          CRSEDESC, GRADE)
```

This table is now in 1NF, since it has no repeating groups. It is not, however, in 2NF, since *STUNAME* is dependent only on *STUNUMB,* which is only a portion of the primary key.

Step 2. Convert the 1NF table to 2NF. First, for each subset of the primary key, start a table with that subset as its key yielding:

```
(STUNUMB,
(CRSENUMB,
(STUNUMB, CRSENUMB,
```

Next, place the rest of the attributes with the minimal collection on which they depend, giving:

```
(STUNUMB, STUNAME, NUMBCRED, ADVNUMB,
          ADVNAME)
(CRSENUMB, CRSEDESC)
(STUNUMB, CRSENUMB, GRADE)
```

Finally, we assign names to each of the newly created tables:

```
STUDENT(STUNUMB, STUNAME, NUMBCRED,
          ADVNUMB, ADVNAME)
COURSE(CRSENUMB, CRSEDESC)
GRADE(STUNUMB, CRSENUMB, GRADE)
```

While these tables are all in 2NF, both *COURSE* and *GRADE* are also in 3NF. The *STUDENT* table is not, however, since it contains a determinant, *ADVNUMB,* that is not a candidate key.

Step 3. Convert the 2NF *STUDENT* table to 3NF by removing the attribute that depends on the determinant *ADVNUMB* and placing it in a separate table:

```
(STUNUMB, STUNAME, NUMBCRED, ADVNUMB)
(ADVNUMB, ADVNAME)
```

Step 4: Name these tables and put the entire collection together, giving:

```
STUDENT(STUNUMB, STUNAME, NUMBCRED,
ADVNUMB)
    ADVISOR(ADVNUMB, ADVNAME)
    COURSE(CRSENUMB, CRSEDESC)
    GRADE(STUNUMB, CRSENUMB, GRADE)
```

SUMMARY

1. Column B is functionally dependent on column (or collection of columns) A if a value of A uniquely determines a value of B at any point in time.
2. The primary key is a column (or collection of columns), A, such that all other columns are functionally dependent on A and no subcollection of the columns in A also has this property.
3. If there is more than one possible choice for the primary key, one of the possibilities is chosen to be *the* primary key. The others are referred to as candidate keys.
4. A relation (table) is in first normal form (1NF) if it does not contain repeating groups.
5. A relation (table) is in second normal form (2NF) if it is in 1NF and if no column that is not a part of the primary key is dependent on only a portion of the primary key.
6. A determinant is any column that functionally determines another column.
7. A relation (table) is in third normal form (3NF) if it is in 2NF and if the only determinants it contains are candidate keys.
8. A collection of relations (tables) that is not in 3NF possesses inherent problems (called update anomalies). Replacing this collection by an equivalent collection of relations (tables) that is in 3NF removes these anomalies. This replacement must be done carefully, following a method like the one proposed in this text. If not, other problems, such as those discussed in this chapter, may very well be introduced.

KEY TERMS

alternate key
Boyce-Codd normal form
   (BCNF)
candidate key
database design
dependency diagram
determinant
first normal form (1NF)

functional dependence
normalization
partial dependency
primary key
repeating group
second normal form (2NF)
third normal form (3NF)
unnormalized relation

EXERCISES

1. Define functional dependence.
2. Give an example of an attribute, A, and another attribute, B, such that B is functionally dependent on A. Give an example of an attribute, C, and an attribute, D, such that D is not functionally dependent on C.
3. Define primary key.
4. Define candidate key.
5. Define first normal form.

6. Define second normal form. What types of problems are encountered in tables that are not in second normal form?

7. Define third normal form. What types of problems are encountered in tables that are not in third normal form?

8. Consider a student table containing student number; student name; student's major department; student's advisor's number; student's advisor's name; student's advisor's office number; student's advisor's phone number; student's number of credits; and student's class standing (freshman, sophomore, and so on). List the functional dependencies that exist, along with the assumptions that would support these dependencies.

9. Using the types of entities found in Henry's system (books, authors, publishers, and so on), create an example of a table that is in 1NF but not in 2NF and an example of a table that is in 2NF but not in 3NF. In each case, justify the answers and show how to convert to the higher forms.

10. Convert the following table to an equivalent collection of tables that is in 3NF.

```
PATIENT(HHNUMB, HHNAME, HHADDR, HHBAL, PATNUMB, PATNAME,
    SERVCODE, SERVDESC, SERVFEE, SERVDATE)
```

This is a table concerning information about patients of a dentist. Each patient belongs to a household. The head of the household is designated as HH in the table. The following dependencies exist in *PATIENT*:

```
PATNUMB --> HHNUMB, HHNAME, HHADDR, HHBAL, PATNAME
HHNUMB --> HHNAME, HHADDR, HHBAL
SERVCODE --> SERVDESC, SERVFEE
PATNUMB, SERVCODE --> SERVDATE
```

11. List the functional dependencies in the following table, subject to the specified conditions. Convert this table to an equivalent collection of tables that are in 3NF.

```
INVOICE(INVNUMB, CUSTNUMB, CUSTNAME, ADDRESS, INVDATE
    PARTNUMB, PARTDESC, UNITPRCE, NUMBSHIP)
```

This table concerns invoice information. For a given invoice (identified by the invoice number) there will be a single customer. The customer's number, name, and address appear on the invoice as well as the invoice date. Also, there may be several different parts appearing on the invoice. For each part that appears, the part number, description, price, and number shipped will be displayed. The price is from the current master price list.

12. Using your knowledge of a college environment, determine the functional dependencies that exist in the following table. After these have been determined, convert this table to an equivalent collection of tables that are in 3NF.

```
STUDENT(STUNUMB, STUNAME, NUMBCRED, ADVNUMB, ADVNAME, DEPTNUMB,
    DEPTNAME, CRSENUMB, CRSEDESC, CRSETERM, GRADE)
```

6

Database Design II: Design Methodology

OBJECTIVES

1. Discuss the general process and goals of database design.
2. Define user views and explain their function.
3. Present a methodology for database design at the information level as well as examples illustrating the use of this methodology.
4. Explain how to produce a pictorial representation of a database design.
5. Explain the process of mapping an information-level design to a design that is appropriate for a relational model system.

6.1 INTRODUCTION

Now that we have learned how to identify and correct bad designs, we will turn our attention to the design process itself; that is, the process of determining the tables and columns that will make up the database and determining the relationships between the various tables.

Database design is often approached as a two-step process. In the first step, a database is designed which satisfies the requirements as cleanly as possible. This step is called **information-level design**, and it is taken *independently* of any particular DBMS that will ultimately be used. In the second step, which is called the **physical-level design**, the information-level design is transformed into a design for the specific DBMS that will be used. Naturally, the characteristics of that DBMS must come into play during this step.

In this text, we will focus on the information-level design process and discuss that portion of the physical-level design process which is geared toward producing a legitimate design for a typical microcomputer DBMS. This approach represents a subset of the design methodology given in [14]. That methodology, which encompasses both the information and physical levels of design, is intended to be used for the design of complex databases that may be implemented on a variety of DBMSs, on mainframes or microcomputers, and where performance can be a very important concern. For the majority of microcomputer applications, the database design process as presented in this text is more than sufficient. If you are interested in delving more deeply into the process, see Chapters 6, 7, and 12 of [14].

6.2 INFORMATION-LEVEL DESIGN

User Views

No matter which approach is adopted with regard to database design, a complete database design that will satisfy all of the requirements can only rarely be a one-step process. Unless the requirements are exceptionally simple, it is usually necessary to subdivide the overall job of database design into smaller tasks. This is often done through the separate consideration of individual pieces of the design problem. In design problems for large organizations, these pieces are often called user views, and we will use the same terminology here. A **user view** is the view of data that is necessary to support the operations of a particular user. For each user view, a database structure to support the view must be designed and then merged into a cumulative design. Each user view, in general, will be much simpler than the total collection of requirements, so working on these individual tasks will be much more manageable than attempting to turn the design of the entire database into one large task.

The General Database Design Methodology

The database design methodology set forth in this text involves representing individual user views, refining them to eliminate any problems, and then merging them into a cumulative design. A "user" could be a person or a group that will use the system, a report the system must produce, or a type of transaction that the system must support. In the last two instances, you might think of the user as the person who will use the report or enter the transaction. In fact, if the same individual required three separate reports, for example, we would probably be better off to consider each of the reports as a separate user view, even though only one "user" was involved, since the smaller the user view, the easier it is to work with.

We now turn to the methodology itself. For each user view, we need to complete the following four steps:

1. Represent the user view as a collection of tables.
2. Normalize these tables.
3. Represent all keys.
4. Merge the result of the previous steps into the design.

We will now examine each of these steps in detail.

6.3 THE METHODOLOGY

Represent the User View as a Collection of Tables

When given a user view or some sort of stated requirement, we must develop a collection of tables that will support it. In some cases, the collection of tables may be obvious to us. Let's suppose, for example, that a given user view involves departments and employees. Let's assume further that each department can employ many employees but that each employee is assigned to exactly one department (a typical restriction). The design

```
DEPT (DEPTNUMB, DEPTNAME, DEPTLOC)
EMPLOYEE (EMPNUMB, EMPNAME, EMPADDR, WAGERATE, SSNUMB, DEPTNUMB)
```

may have naturally occurred to you and is an appropriate design. You will undoubtedly find that the more designs you have done, the easier it will be for you to develop such a collection without resorting to any special procedure. The real question is, What procedure should be followed if a correct design is not so obvious? In this case, we can take the following four steps:

Step 1. Determine the entities involved and create a separate table for each type of entity. At this point, you do not need to do anything more than give the table a name. For example, if a user view involves departments and employees, we can create a *DEPT* table and an *EMPLOYEE* table. At this point, we will write down something like this:

```
DEPT (
EMPLOYEE (
```

That is, we will write down the name of a table and a left parenthesis, *and that is all*. Later steps will fill in the attributes in these tables.

Step 2. Determine the primary key for each of these tables. This will fill in one or two attributes (depending on how many attributes make up the primary key). Other attributes will not be filled in until a later step. It may seem strange, but even though we have yet to determine the attributes in the table, we can usually determine the primary key. For example, the primary key to an *EMPLOYEE* table will probably be the employee number, and the primary key to a *DEPT* table will probably be the department number.

The primary key is the unique identifier, so the essential question here is, What does it take to uniquely identify an employee or a department? Even if we are in the process of trying to automate a system that was previously manual, some unique identifier can still usually be found in the manual system. If not, it is probably time to assign one. Let's say, for example, that in a particular manual system customers did not have numbers. The customer base was small enough that the organization felt they were not needed. Now is a good time to assign them, however, since the company is computerizing. These numbers would then be the unique identifier we are seeking.

Now let's add these primary keys to what we have written down already. At this point, we will have something like the following:

```
DEPT (DEPTNUMB,
EMPLOYEE (EMPNUMB,
```

That is, we will have the name of the table and the primary key, but that is all. Later steps will fill in the other attributes.

Step 3. Determine the properties for each of these entities. We can look at the user requirements and then determine the other properties of each entity which are required. These properties, along with the key identified in step 2, will become attributes in the appropriate tables. For example, an employee entity may require *EMPNAME*, *EMPADDR*, *WAGERATE*, and *SSNUMB* (social security number). The department entity may require *DEPTNAME* (department name) and *DEPTLOC* (department location). Adding these to what is already in place would produce the following:

```
DEPT (DEPTNUMB, DEPTNAME, DEPTLOC)
EMPLOYEE (EMPNUMB, EMPNAME, EMPADDR, WAGERATE, SSNUMB)
```

Step 4. Determine relationships among the entities. The basic relationships are one-to-many, many-to-many, and one-to-one. We will now see how to handle each of these types of relationships.

One-to-many. A one-to-many relationship is implemented by including the primary key of the "one" table as a foreign key in the "many" table. Let's suppose, for example, that each employee is assigned to a single department but that a department can have many employees. Thus *one* department is related to *many* employees. In this case, we would include the primary key of the *DEPT* table (the "one") as a foreign key in the *EMPLOYEE* table (the "many"). Thus, the tables would now look like this:

```
DEPT (DEPTNUMB, DEPTNAME, DEPTLOC)
EMPLOYEE (EMPNUMB, EMPNAME, EMPADDR, WAGERATE, SSNUMB, DEPTNUMB)
```

Many-to-many. A many-to-many relationship is implemented by creating a new table whose key is the combination of the keys of the original tables. Let's suppose that each employee can be assigned to multiple departments and that each department can have many employees. In this case, we would create a new table whose primary key would be the combination of *EMPNUMB* and *DEPTNUMB*. Since the new table represents the fact that an employee *works in* a department, we might choose to call it *WORKSIN*, in which case the collection of tables is as follows:

```
DEPT (DEPTNUMB, DEPTNAME, DEPTLOC)
EMPLOYEE (EMPNUMB, EMPNAME, EMPADDR, WAGERATE, SSNUMB)
WORKSIN (EMPNUMB, DEPTNUMB)
```

In some situations, no other attributes will be required in the new table. The other attributes in the *WORKSIN* table would be those attributes which depended on both the employee and the department, if such attributes existed. One possibility, for example, would be the date when the employee was first assigned to the department, since it depends on *both* the employee *and* the department.

One-to-one. If each employee is assigned to a single department and each department consists of only one employee, the relationship between employees and departments is one-to-one. The simplest way for us to implement a one-to-one relationship is to treat it as a one-to-many relationship. But which is the "one" part of the relationship and which is the "many" part? Sometimes looking to the future helps. For instance, in the example we are discussing, we might ask, If the relationship changes in the future, is it more likely that one employee will be assigned to many departments or that one department may consist of several employees rather than just one? If we feel, for example, that it is more likely that a department would be allowed to contain more than one employee, we would make *EMPLOYEE* the "many" part of the relationship. If the answer is that both things might very well happen, we might even treat the relationship as many-to-many. If neither change were likely to occur, we could actually resort to flipping a coin in order to choose the "many" part of the relationship.

Normalize These Tables

Normalize each table, with the target being third normal form.

Represent All Keys

Identify all keys. The types of keys we must identify are primary keys, alternate keys, secondary keys, and foreign keys.

1. *Primary:* The primary key has already been determined in the earlier steps.
2. *Alternate:* An **alternate key** is an attribute or collection of attributes that could have been chosen as a primary key but was not. It is not common to have alternate keys; but if they do exist, and if the system is to enforce their uniqueness, they should be so noted.
3. *Secondary:* If there are any **secondary keys** (attributes that are of interest strictly for the purpose of retrieval), they should be represented at this point. If a user were to indicate, for example, that rapidly retrieving an employee on the basis of his or her name was important, we would designate *EMPNAME* as a secondary key.
4. *Foreign:* This is in many ways the most important category, since it is through **foreign keys** that relationships are established and that certain types of integrity constraints are enforced in the database. Remember that a foreign key is an attribute (or collection of attributes) in one table that is required to either match

the value of the primary key for some row in another table or be null. (This is the property called **referential integrity**.) Consider, for example, the following tables:

```
DEPT (DEPTNUMB, DEPTNAME, DEPTLOC)
EMPLOYEE (EMPNUMB, EMPNAME, EMPADDR, WAGERATE, SSNUMB, DEPTNUMB)
```

As before, *DEPTNUMB* in the *EMPLOYEE* table indicates the department to which the employee is assigned. We say that *DEPTNUMB* in the *EMPLOYEE* table is a foreign key that *identifies DEPT*. Thus, the number in this attribute on any row in the *EMPLOYEE* table must either be the number of a department that is already in the database, or be null. (Null would indicate that, for whatever reason, the employee is not assigned to a department.)

Database Design Language (DBDL)

We need a mechanism for representing the tables and keys together with the restrictions previously discussed. The standard mechanism for representing tables is fine but it does not go far enough. There is no routine way to represent alternate, secondary, or foreign keys, nor is there a way of representing foreign key restrictions. There is no way of indicating that a given field or attribute can accept null values. Since the methodology is based on the relational model, however, it is desirable to represent tables with the standard method. We will add additional features capable of representing additional information. The end result is **Database Design Language** (or **DBDL**).

Figure 6.1a shows sample DBDL documentation for the *EMPLOYEE* table. In DBDL, tables and their primary keys are represented in the usual manner. Any field that is allowed to be null, such as the *EMPADDR* attribute in the *EMPLOYEE* table, is followed by an asterisk. Underneath the table, the various types of keys are listed. Each is preceded by an abbreviation indicating the type of key (AK – alternate key, SK – secondary key, FK – foreign key). It is sufficient to list the attribute or collection of attributes that forms an alternate or secondary key. In the case of foreign keys, however, we must also represent the table that is identified by the foreign key; i.e., the table whose primary key the foreign key must match. This is accomplished in DBDL by following the foreign key with an arrow pointing to the table that the foreign key identifies.

Figure 6.1a

DBDL for *EMPLOYEE* relation

```
EMPLOYEE (EMPNUMB, EMPNAME, EMPADDR*, SSNUMB, DEPTNUMB, ...)
         AK    SSNUMB
         SK    EMPNAME
         FK    DEPTNUMB --> DEPT
```

Figure 6.1b summarizes the details of DBDL. Examples of DBDL will be presented throughout this chapter. The only feature of DBDL not listed is actually more of a tip than a rule. When several tables are listed, a table containing a foreign key should be listed after the table that the foreign key identifies, if possible.

Figure 6.1b

Summary of DBDL

DBDL (Database Design Language)

1. Relations, attributes, and primary keys are represented in the usual way.
2. Attributes that are allowed to be null are followed by an asterisk.
3. Alternate keys are identified by the letters AK followed by the attribute(s) that comprise the alternate key.
4. Secondary keys are identified by the letters SK followed by the attribute(s) that comprise the secondary key.
5. Foreign keys are identified by the letters FK followed by the attribute(s) that comprise the foreign key. Foreign keys are followed by an arrow pointing to the relation identified by the foreign key.

In the example shown in Figure 6.1a, we are saying that there is a table called *EMPLOYEE*, consisting of fields *EMPNUMB*, *EMPNAME*, *EMPADDR*, *SSNUMB* (social security number), *DEPTNUMB*, and so on. The *EMPADDR* field is the only one that can accept null values. The primary key is *EMPNUMB*. Another possible key is *SSNUMB*. We are interested in being able to retrieve information efficiently, based on the employee's name, so we have designated *EMPNAME* as a secondary key. The *DEPTNUMB* is a foreign key identifying the department to which the employee is assigned (it identifies the appropriate department in the *DEPT* table).

A Pictorial Representation of the Database

For many people, a pictorial representation of the structure of the database is quite useful. (As the old saying goes, "A picture is worth a thousand words.") Fortunately, there is an easy procedure for including a diagram representing the database structure in DBDL. The type of diagram we will use is often called a data structure diagram. The procedure for constructing such a diagram from tables represented in DBDL is as follows:

1. Draw a rectangle for each table in the DBDL design. Label the rectangle with the name of the corresponding table.
2. For each foreign key, draw an arrow from the rectangle that corresponds to the table being identified to the rectangle that corresponds to the table containing the foreign key.
3. In the rare event that you have two arrows joining the same two rectangles, label the arrows with names that are indicative of the meaning of the relationships represented by the arrows.
4. If the diagram you have drawn is cluttered or messy, redraw the diagram. If possible, avoid crossing arrows, since this makes the diagram more difficult to understand.

Figure 6.2 shows the DBDL from Figure 6.1a together with a corresponding data structure diagram. Notice that there is a *DEPT* rectangle and an *EMPLOYEE* rectangle. Further, since the *EMPLOYEE* table contains a foreign key identifying the *DEPT* table, there is an arrow from *DEPT* to *EMPLOYEE*. This arrow visually emphasizes the relationship between departments and employees. Such arrows represent one-to-many relationships (*one* department to *many* employees) with the arrow pointing to the "many" part of the relationship.

Figure 6.2

DBDL with data structure diagram

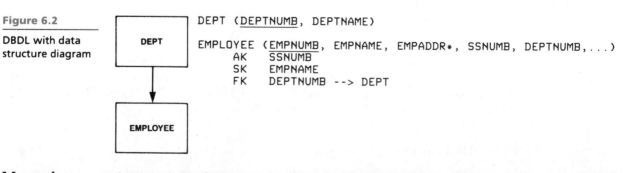

```
DEPT (DEPTNUMB, DEPTNAME)

EMPLOYEE (EMPNUMB, EMPNAME, EMPADDR*, SSNUMB, DEPTNUMB,...)
      AK    SSNUMB
      SK    EMPNAME
      FK    DEPTNUMB --> DEPT
```

Merge the Result into the Design

As soon as we have completed steps 1 through 3 for a given user view, we can merge these results into the overall design. If the view on which we have been working happens to be the first user view, then the cumulative design will be identical to the design for this first user. Otherwise, we add all the tables for this user to those which are currently in the cumulative design. We combine tables that have the same primary key to form a new table. This table has the same primary key as those tables which have been combined.

The new tables also contain all the attributes from both tables. In the case of duplicate attributes, we remove all but one copy of the attribute. For example, if the cumulative collection already contained the following:

```
EMPLOYEE (EMPNUMB, EMPNAME, WAGERATE, SSNUMB, DEPTNUMB)
```

and the user view just completed contained the following:

```
EMPLOYEE (EMPNUMB, EMPNAME, EMPADDR)
```

then the two tables would be combined, since they would have the same primary key. All of the attributes from both tables would appear in the new table, but without duplicates. Thus, *EMPNAME* would appear only once, even though it is in each of the individual tables. The result would be the following:

```
EMPLOYEE (EMPNUMB, EMPNAME, WAGERATE, SSNUMB, DEPTNUMB, EMPADDR)
```

If we wanted to, we could reorder the attributes at this point. We might feel, for example, that placing *EMPADDR* immediately after *EMPNAME* would put it in a more natural position. This would give the following:

```
EMPLOYEE (EMPNUMB, EMPNAME, EMPADDR, WAGERATE, SSNUMB, DEPTNUMB)
```

We would then check the new design to ensure that it was still in third normal form. If it was not, we would convert it to 3NF before proceeding.

The process, which is summarized in Figure 6.3, is repeated for each user view until all user views have been examined. At that point, the design is reviewed in order to resolve any problems that may remain and to ensure that the needs of all individual users can indeed be met. Once this has been done, the information-level design is complete.

Figure 6.3

Information-level design methodology

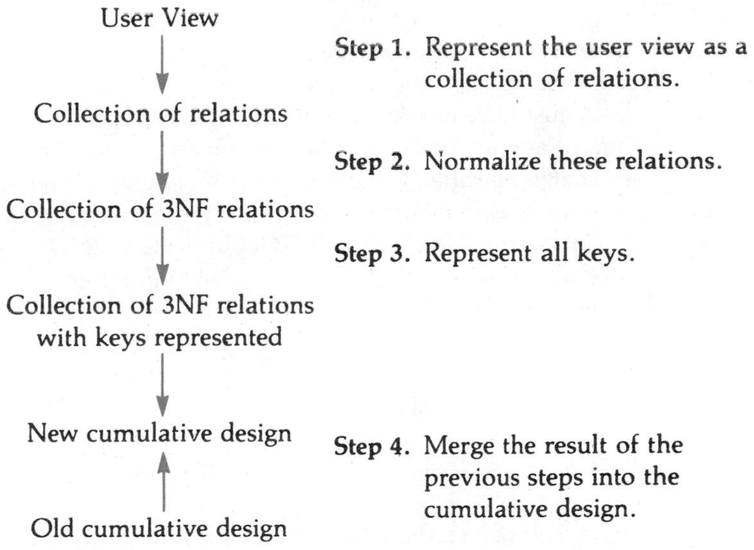

User View

↓

Collection of relations

↓

Collection of 3NF relations

↓

Collection of 3NF relations with keys represented

↓

New cumulative design

↑

Old cumulative design

Step 1. Represent the user view as a collection of relations.

Step 2. Normalize these relations.

Step 3. Represent all keys.

Step 4. Merge the result of the previous steps into the cumulative design.

6.4 DATABASE DESIGN EXAMPLES

Let's now look at some examples of database design.

Example 1: For an initial example of the design methodology, let's complete an information-level design for a database that must satisfy the following constraints and requirements.

1. For a sales rep, store the sales rep's number, name, address, total commission and commission rate.
2. For a customer, store the customer's number, name, address, balance, and credit limit. In addition, store the number and name of the sales rep who represents this customer. Upon further checking with the user we determine that a sales rep can represent many customers but a customer must have exactly one sales rep (i.e., a customer *must have* a sales rep and cannot have more than *one*.)
3. For a part, store the part's number, description, units on hand, item class, the number of the warehouse in which the part is located, and the price.
4. For an order, store the order number, order date, the number, name, and address of the customer who placed the order, and the number of the sales rep who represents that customer. In addition, for each line item within the order, store the part number and description, the number of the part that was ordered, and the quoted price. The following information has also been obtained from the user:
   a. Each order must be placed by a customer who is already in the customer file.
   b. There is only one customer per order.
   c. On a given order, there is at most one line item for a given part. For example, part BT04 cannot appear on several lines within the same order.
   d. The quoted price may be the same as the current price in the part master file, but it need not be. This allows the enterprise the flexibility to sell the same parts to different customers for different prices. It also allows you to change the basic price for a part without necessarily affecting other orders that are currently on file.

What are the user views in the preceding example? In particular, how should the design proceed if we are given requirements that are not specifically stated in the form of user views? We might actually be lucky enough to be confronted with a series of well-thought-out user views in a form that can readily be merged into our design. On the other hand, we might only be given a set of requirements like the set we have encountered in this example. Or we might be given a list of reports and updates that a system must support. If we happen to be given the job of interviewing users and documenting their needs as a preliminary to the design process, we can make sure that their views are specified in a form that will be easy to work with when the design process starts. On the other hand, we may just have to take this information as we get it.

If the user views are not spelled out as user views per se, then we should consider each requirement that is specified to be a user view. Thus each report or update transaction that the system must support, as well as any other requirement, such as any of those just stated, can be considered an individual user view. In fact, even if the requirements are presented as user views, we may wish to split up a user view that is particularly complex into smaller pieces and consider each piece a user view for the design process.

Let us now proceed with the example.

User View 1: This requirement, or user view, poses no particular difficulty. Only one table is required to support this view:

SLSREP (<u>SLSRNUMB</u>, SLSRNAME, SLSRADDR, TOTCOMM, COMMRATE)

This table is in 3NF. Since there are no foreign, alternate, or secondary keys, the DBDL representation of the table is precisely the same as the relational model representation.

Notice that we have assumed that the sales rep's number (*SLSRNUMB*) is the primary key to the table. This is a fairly reasonable assumption. But since this information was not given in the first requirement, we would need to verify its accuracy with the user.

In each of the following requirements, we shall assume that the obvious attribute (customer number, part number, and order number) is the primary key. Since this is the first user view, the "merge" step of the design methodology will produce a cumulative design consisting of this one table (see Figure 6.4).

Figure 6.4

Cumulative design
after first user view

```
┌─────────────┐        SLSREP (SLSRNUMB, SLSRNAME, SLSRADDR, TOTCOMM, COMMRATE)
│   SLSREP    │
└─────────────┘
```

User View 2: Because the first user view was relatively simple, we were able to come up with the necessary table without having to go through the steps mentioned in the discussion of the design methodology. The second user view is a little more complicated, however, so let's use the steps suggested earlier to determine the tables. (If you've already spotted what the tables should be, you have a natural feel for the process. If so, please be patient while we work through the process.)

We'll take two different approaches to this requirement so that we can see how they can both lead to the same result. The only difference between the two approaches concerns the entities that we initially identify. In the first approach, suppose we identify two entities, *sales reps* and *customers*. We would then begin with the two following tables:

```
SLSREP (
CUSTOMER (
```

After determining the unique identifiers, we add the primary keys, which would give:

```
SLSREP (SLSRNUMB,
CUSTOMER (CUSTNUMB,
```

Adding attributes for the properties of each of these entities would yield:

```
SLSREP (SLSRNUMB, SLSRNAME
CUSTOMER (CUSTNUMB, CUSTNAME, ADDRESS, BALANCE, CREDLIM
```

Finally, we would deal with the relationship: *one* sales rep is related to *many* customers. To implement this one-to-many relationship, we would include the key of the "one" table in the "many" table as a foreign key. In this case, we would include *SLSRNUMB* in the *CUSTOMER* table. Thus, we would have the following:

```
SLSREP (SLSRNUMB, SLSRNAME)
CUSTOMER (CUSTNUMB, CUSTNAME, ADDRESS, BALANCE, CREDLIM,
          SLSRNUMB)
```

Both tables are in 3NF, so we can move on to representing the keys. Before doing that, however, let's investigate another approach that could have been used to determine the tables.

Suppose we didn't realize that there were really two entities and thought there was only a single entity, *customers*. We would thus begin only the single table as follows:

```
CUSTOMER (
```

Adding the unique identifier as the primary key would give this:

```
CUSTOMER (CUSTNUMB,
```

Finally, adding the other properties as additional attributes would yield:

```
CUSTOMER (CUSTNUMB, CUSTNAME, ADDRESS, BALANCE, CREDLIM,
          SLSRNUMB, SLSRNAME)
```

A problem appears, however, when we examine the functional dependencies that exist in *CUSTOMER*. *CUSTNUMB* determines all the other fields, as it should. But *SLSRNUMB* determines *SLSRNAME*, yet *SLSRNUMB* is not a candidate key. This table, which is in 2NF, since no attribute depends on a portion of the key, is not in 3NF. Thus, converting to 3NF would produce the following two tables:

```
CUSTOMER (CUSTNUMB, CUSTNAME, ADDRESS, BALANCE, CREDLIM,
          SLSRNUMB)
SLSREP (SLSRNUMB, SLSRNAME)
```

Note that these are precisely the same tables that we determined with the other approach. It just took us a little longer to get there.

It is these two tables that we merge into the design. Besides the obvious primary keys, *CUSTNUMB* for *CUSTOMER* and *SLSRNUMB* for *SLSREP*, the *CUSTOMER* table now contains a foreign key, *SLSRNUMB*.

There are no alternate keys, nor did the requirements state anything that would lead to a secondary key. If there were a requirement to retrieve the customer based on his or her name, for example, we would probably choose to make *CUSTNAME* a secondary key. (Since names are not unique, *CUSTNAME* is not an alternate key.)

At this point, we could represent the table *SLSREP* in DBDL in preparation for merging this collection of tables into the collection we already have. Looking ahead, however, we see that since this table has the same primary key as the table *SLSREP* from the first user view, the two tables will be merged. A single table will be formed that has the common key *SLSRNUMB* as its primary key and that contains all of the other attributes from both tables without duplication. For this second user view, the only attribute in *SLSREP* besides the primary key is *SLSRNAME*. This attribute is the same as the attribute called *SLSRNAME* already present in *SLSREP* from the first user view. Thus, nothing will be added to the *SLSREP* table that is already in place. The cumulative design now contains the two tables *SLSREP* and *CUSTOMER*, as shown in Figure 6.5.

Figure 6.5

Cumulative design after second user view

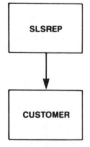

```
SLSREP (SLSRNUMB, SLSRNAME, SLSRADDR, TOTCOMM, COMMRATE)

CUSTOMER (CUSTNUMB, CUSTNAME, ADDRESS, BALANCE, CREDLIM, SLSRNUMB)
    FK   SLSRNUMB --> SLSREP
```

User View 3: Like the first user view, this one poses no special problems. Only one table is required to support it:

```
PART (PARTNUMB, PARTDESC, UNONHAND, ITEMCLSS, WREHSENM, UNITPRCE)
```

Figure 6.6

Cumulative design after third user view

This table is in 3NF. The DBDL representation is identical to the relational model representation.

Since *PARTNUMB* is not the primary key of any table we have already encountered, merging this table into the cumulative design produces a design with the three tables *SLSREP*, *CUSTOMER*, and *PART* (see Figure 6.6).

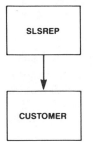

```
SLSREP (SLSRNUMB, SLSRNAME, SLSRADDR, TOTCOMM, COMMRATE)

CUSTOMER (CUSTNUMB, CUSTNAME, ADDRESS, BALANCE, CREDLIM, SLSRNUMB)
    FK   SLSRNUMB --> SLSREP

PART (PARTNUMB, PARTDESC, UNUNHAND, ITEMCLSS, WREHSENM, UNITPRCE)
```

User View 4: This user view is a bit more complicated, and we could approach it in several ways. Suppose we felt that only a single entity was being mentioned, namely *orders*. In that case, we would create a single table, as follows:

```
ORDERS (
```

Since orders are uniquely identified by order numbers, we would add *ORDNUMB* as the primary key, giving:

```
ORDERS (ORDNUMB,
```

Examining the various properties of an order, such as the date, the customer number, and so on, as listed in the requirement, we would add appropriate attributes, giving.

```
ORDERS (ORDNUMB, ORDDTE, CUSTNUMB, CUSTNAME, ADDRESS, SLSRNUMB,
```

What about the fact that we are supposed to store the part number, description, number ordered, and quoted price for each order line on this order? One way of doing this would be to include all these attributes within the *ORDERS* table as a repeating group (since there can be many order lines on an order). This would yield:

```
ORDERS (ORDNUMB, ORDDTE, CUSTNUMB, CUSTNAME, ADDRESS, SLSRNUMB,
        PARTNUMB, PARTDESC, NUMBORD, QUOTPRCE)
```

At this point, we have a table that does contain all the necessary attributes. Now we must convert this table to an equivalent collection of tables that are in 3NF. Since this table is not even in 1NF, we would remove the repeating group and expand the key to produce the following:

```
ORDERS (ORDNUMB, ORDDTE, CUSTNUMB, CUSTNAME, ADDRESS, SLSRNUMB,
        PARTNUMB, PARTDESC, NUMBORD, QUOTPRCE)
```

In the new *ORDERS* table, we have the following functional dependencies:

```
ORDNUMB --> ORDDTE, CUSTNUMB, CUSTNAME, ADDRESS, SLSRNUMB
CUSTNUMB --> CUSTNAME, ADDRESS, SLSRNUMB
PARTNUMB --> PARTDESC
ORDNUMB, PARTNUMB --> NUMBORD, QUOTPRCE
```

From the discussion of the quoted price in the statement of the requirement, it should be noted that quoted price does indeed depend on *both* the order number and the part number, not on the part number alone. Since some attributes depend on only a portion of the primary key, the *ORDERS* table is not in 2NF. Converting to 2NF would yield the following:

```
ORDERS (ORDNUMB, ORDDTE, CUSTNUMB, CUSTNAME, ADDRESS, SLSRNUMB)
PART (PARTNUMB, PARTDESC)
ORDLNE (ORDNUMB, PARTNUMB, NUMBORD, QUOTPRCE)
```

The tables *PART* and *ORDLNE* are in 3NF. The *ORDERS* table is not in 3NF, since *CUSTNUMB* determines *NAME, ADDRESS,* and *SLSRNUMB*, but *CUSTNUMB* is not a candidate key. Converting the *ORDERS* table to 3NF and leaving the other tables untouched would produce the following design for this requirement.

```
ORDERS (ORDNUMB, ORDDTE, CUSTNUMB)
CUSTOMER (CUSTNUMB, CUSTNAME, ADDRESS, SLSRNUMB)
PART (PARTNUMB, PARTDESC)
ORDLNE (ORDNUMB, PARTNUMB, NUMBORD, QUOTPRCE)
```

This is the collection of tables that will be represented in DBDL and then merged into the cumulative design. Again, however, we can look ahead and see that *CUSTOMER* will be merged with the existing *CUSTOMER* table, and *PART* will be merged with the existing *PART* table. In neither case will anything new be added to the *CUSTOMER* and *PART* tables already in place, so the *CUSTOMER* and *PART* tables for this user view will not affect the overall design. The representation for this user view in DBDL is shown in Figure 6.7.

Figure 6.7

DBDL for fourth user view

```
CUSTOMER (CUSTNUMB, CUSTNAME, ADDRESS, SLSRNUMB)

PART (PARTNUMB, PARTDESC)

ORDERS (ORDNUMB, ORDDTE, CUSTNUMB)
    FK   CUSTNUMB --> CUSTOMER

ORDLNE (ORDNUMB, PARTNUMB, NUMBORD, QUOTPRCE)
    FK   ORDNUMB --> ORDERS
    FK   PARTNUMB --> PART
```

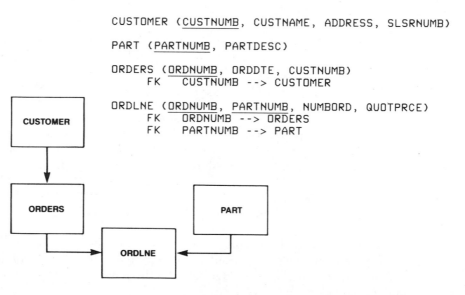

At this point, we have completed the process for each user. We should now review the design to make sure that it will cleanly fulfill all of the requirements. If problems are encountered or new information comes to light, the design must be modified accordingly. Based on the assumption that we do not have to further modify the design here, the final information-level design is shown in Figure 6.8, which is the result of merging the design for the fourth user view into the cumulative design.

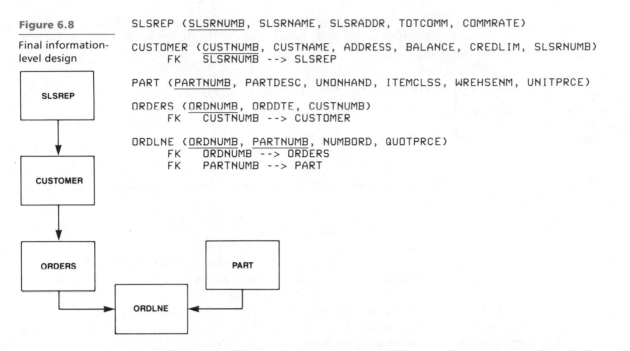

Figure 6.8

Final information-level design

```
SLSREP (SLSRNUMB, SLSRNAME, SLSRADDR, TOTCOMM, COMMRATE)

CUSTOMER (CUSTNUMB, CUSTNAME, ADDRESS, BALANCE, CREDLIM, SLSRNUMB)
     FK    SLSRNUMB --> SLSREP

PART (PARTNUMB, PARTDESC, UNONHAND, ITEMCLSS, WREHSENM, UNITPRCE)

ORDERS (ORDNUMB, ORDDTE, CUSTNUMB)
     FK    CUSTNUMB --> CUSTOMER

ORDLNE (ORDNUMB, PARTNUMB, NUMBORD, QUOTPRCE)
     FK    ORDNUMB --> ORDERS
     FK    PARTNUMB --> PART
```

Example 2: We will now design a database for Henry. Henry wants to keep information on books, authors, publishers, and branches. The only user is Henry, but we don't want to treat the whole project as a single user view: so let's assume we've asked Henry for all the reports the system is to produce, and we will treat each one as a user view. Suppose that Henry has given us the following requirements:

1. For each publisher, list the publisher code, the name, and the city in which the publisher is located.
2. For each branch, list the number, the name, the location, and the number of employees.
3. For each book, list its code, title, the code and name of the publisher, the price, and whether or not it is paperback.
4. For each book, list its code, title, type, and price. In addition, list the number and name of each of the authors of the book. (**Note:** If there is more than one author, they must be listed in the order in which they are listed on the book. This may or may not be alphabetically.)
5. For each branch, list the number and name. In addition, list the code and title of each book currently in the branch as well as the number of units of the book the branch currently has.
6. For each book, list the code and title. In addition, for each branch currently having the book in stock, list the number and name of each branch along with the number of copies available.

With these six reports as the user views, let's move on to the design of Henry's database.

User View 1: The only entity in this user view is *publshr*. The table to support it is as follows:

```
PUBLSHR (PUBCODE, PUBNAME, PUBCITY)
```

This table is in 3NF. The primary key is *PUBCODE*. There are no alternate or foreign keys. Let's assume Henry wants to be able to access a publisher rapidly on the basis of its name. Then we will make *PUBNAME* a secondary key.

Since this is the first user view, there is no previous cumulative design. So at this point the new cumulative design will consist solely of the design for this user view. It is shown in Figure 6.9.

Figure 6.9

DBDL for *BOOK* database after first requirement

PUBLSHR (PUBCODE, PUBNAME, PUBCITY)
    SK PUBNAME

User View 2: The only entity in this user view is *branch*. The table to support it is as follows:

BRANCH (BRNUMB, BRNAME, BRLOC, NUMEMP)

This table is also in 3NF. The primary key is *BRNUMB,* and there are no alternate or foreign keys. Let's assume Henry wants to be able to access a branch rapidly on the basis of its name. Thus we will make *BRNAME* a secondary key.

Since no table in the cumulative design has *BRNUMB* as its primary key, this table will simply be added to the collection of tables in the cumulative design during the merge step. The result is shown in Figure 6.10.

Figure 6.10

DBDL for *BOOK* database after second requirement

PUBLSHR (PUBCODE, PUBNAME, PUBCITY)
    SK PUBNAME

BRANCH (BRNUMB, BRNAME, BRLOC, NUMEMP)
    SK BRNAME

User View 3: There are two entities here, publishers and books, and a one-to-many relationship between them. This leads to the following:

PUBLSHR (PUBCODE, PUBNAME)
BOOK (BKCODE, BKTITLE, PUBCODE, BKPRICE, PB)

where *PUBCODE* in *BOOK* is a foreign key identifying the publisher. Merging these tables with those which are already in place does not add any new attributes to the *PUBLSHR* table but adds the *BOOK* table to the cumulative design. The result of the merge is shown in Figure 6.11.

Figure 6.11

DBDL for *BOOK* database after third requirement

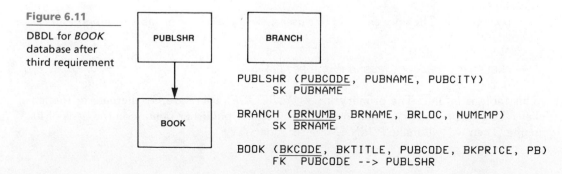

PUBLSHR (PUBCODE, PUBNAME, PUBCITY)
    SK PUBNAME

BRANCH (BRNUMB, BRNAME, BRLOC, NUMEMP)
    SK BRNAME

BOOK (BKCODE, BKTITLE, PUBCODE, BKPRICE, PB)
    FK PUBCODE --> PUBLSHR

User View 4: There are two entities in this user view, books and authors. The relationship between them is many-to-many (an author can write many books and a book can have many authors). Creating tables for each entity and the relationship gives:

```
AUTHOR (AUTHNUMB, AUTHNAME)
BOOK (BKCODE, BKTITLE, BKTYPE, BKPRICE)
WROTE (BKCODE, AUTHNUMB
```

(Since the last table represents the fact that an author *wrote* a particular book, we will call the table *WROTE*).

In this user view, we need to be able to list the authors for a book in the appropriate order. To accomplish this, we will add a sequence number column to the last table. This completes the tables for this user view, which are:

```
AUTHOR (AUTHNUMB, AUTHNAME)
BOOK (BKCODE, BKTITLE, BKTYPE, BKPRICE)
WROTE (BKCODE, AUTHNUMB, SEQNO)
```

The *AUTHOR* table is new. Merging the *BOOK* table adds one additional column, *BKTYPE*. The *WROTE* table is new. The result of the merge step is shown in Figure 6.12.

Figure 6.12

DBDL for *BOOK* database after fourth requirement

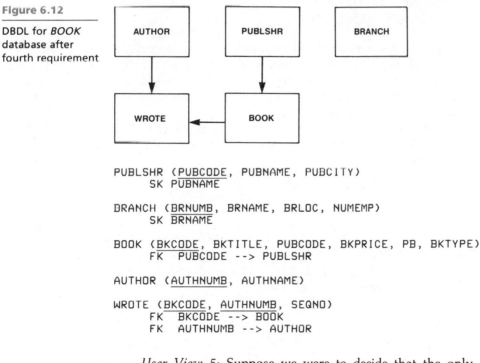

```
PUBLSHR (PUBCODE, PUBNAME, PUBCITY)
    SK  PUBNAME

BRANCH (BRNUMB, BRNAME, BRLOC, NUMEMP)
    SK  BRNAME

BOOK (BKCODE, BKTITLE, PUBCODE, BKPRICE, PB, BKTYPE)
    FK  PUBCODE --> PUBLSHR

AUTHOR (AUTHNUMB, AUTHNAME)

WROTE (BKCODE, AUTHNUMB, SEQNO)
    FK  BKCODE --> BOOK
    FK  AUTHNUMB --> AUTHOR
```

User View 5: Suppose we were to decide that the only entity mentioned in this requirement was *branch*. We would then create this table:

```
BRANCH (
```

We would then add the primary key, *BRNUMB*, producing the following:

```
BRANCH (BRNUMB,
```

The other properties include the branch name as well as the book code, book title, and number of units on hand. Since a branch will have several books, the last three columns will form a repeating group. We thus have the following:

BRANCH (<u>BRNUMB</u>, BRNAME, <u>BKCODE, BKTITLE, OH</u>)

We convert this table to 1NF by removing the repeating group and expanding the key. This gives:

BRANCH (<u>BRNUMB</u>, BRNAME, <u>BKCODE</u>, BKTITLE, OH)

In this table, we have the following functional dependencies:

BRNUMB --> BRNAME
BKCODE --> BKTITLE
BRNUMB, BKCODE --> OH

The table is not in 2NF, since some attributes depend on just a portion of the key. Converting to 2NF gives:

BRANCH (<u>BRNUMB</u>, BRNAME)
BOOK (<u>BKCODE</u>, BKTITLE)
INVENT (<u>BRNUMB</u>, <u>BKCODE</u>, OH)

The primary keys are indicated. We call the final table *INVENT*, since it effectively represents each branch's inventory. In the *INVENT* table, *BRNUMB* is a foreign key that identifies *BOOK*, and *BKCODE* is a foreign key that identifies *BOOK*. In other words, in order for a row to exist in the *INVENT* table, *both* the branch number *and* the book code must already be in the database.

The *BRANCH* table will merge with the existing *BRANCH* table without adding anything new. Similarly, the *BOOK* table will not add anything new to the existing *BOOK* table. The *INVENT* table is totally new and will appear as part of the new cumulative design, which is given in Figure 6.13.

Figure 6.13

DBDL for *BOOK* database after fifth requirement

PUBLSHR (<u>PUBCODE</u>, PUBNAME, PUBCITY)
     SK <u>PUBNAME</u>

BRANCH (<u>BRNUMB</u>, BRNAME, BRLOC, NUMEMP)
     SK <u>BRNAME</u>

BOOK (<u>BKCODE</u>, BKTITLE, PUBCODE, BKPRICE, PB, BKTYPE)
     FK PUBCODE --> PUBLSHR

AUTHOR (<u>AUTHNUMB</u>, AUTHNAME)

WROTE (<u>BKCODE</u>, <u>AUTHNUMB</u>, SEQNO)
     FK BKCODE --> BOOK
     FK AUTHNUMB --> AUTHOR

INVENT (<u>BRNUMB</u>, <u>BKCODE</u>, OH)
     FK BRNUMB --> BRANCH
     FK BKCODE --> BOOK

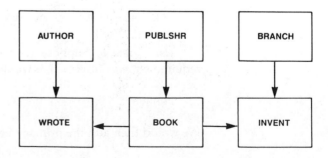

Question: How would the design for this user view have turned out if we had started out with two entities, *branch* and *book*, instead of just the single entity *branch*?

Answer: In the first step, we would have these two tables:

BRANCH (
BOOK (

Adding the primary keys would give:

BRANCH (BRNUMB,
BOOK (BKCODE,

Filling in the other attributes would give:

BRANCH (BRNUMB, BRNAME)
BOOK (BKCODE, BKTITLE)

Finally, we have to implement the relationship between *BRANCH* and *BOOK*. Since a branch can have many books and a book can be in stock at many branches, the relationship is many-to-many. To implement a many-to-many relationship, we add a new table whose primary key is the combination of the primary keys of the other tables. Doing this, we produce the following:

BRANCH (BRNUMB, BRNAME)
BOOK (BKCODE, BKTITLE)
INVENT (BRNUMB, BKCODE

Finally, we add any column that depends on both *BRNUMB* and *BKCODE* to the *INVENT* table, giving

BRANCH (BRNUMB, BRNAME)
BOOK (BKCODE, BKTITLE)
INVENT (BRNUMB, BKCODE, OH)

Thus we end up with exactly the same collection of tables, which illustrates a point made earlier: there's more than one way of arriving at a correct result.

User View 6: This user view leads to precisely the same set of tables that were created for User View 5.

We have now reached the end of the requirements, and the design shown in Figure 6.13 represents the complete information-level design. You should take a moment to review each of the requirements to make sure they can all be satisfied.

(**Note:** If you compare this design with the *BOOK* database we have seen earlier in the text, you will see a slight difference in the order of some of the columns. In theory it doesn't make any difference. In practice, however, we sometimes rearrange the columns when we are done for convenience. If, for example, you execute a SQL SELECT using the star (*), you will see the columns *in the order in which they occur in the table.* Thus, if we have a particular order we prefer, we will often be sure the columns occur in that order.)

6.5 PHYSICAL-LEVEL DESIGN

Once the information-level design is complete, we are ready to begin producing the specific design that will be implemented with some typical microcomputer DBMS. This is part of the overall process called physical-level design. For further information on the full physical-level design process, see [14].

Since most microcomputer DBMSs are relational (at least they claim to be), and since our final information-level design is presented in a relational format, the basic job of producing the design for the chosen DBMS is not difficult. We simply use the same tables and columns. (At this point, we do need to supply format details, of course, like the fact that *CUSTNUMB* is a three-digit number, but again, this is not difficult.)

If the DBMS happened to support primary, candidate, secondary, and foreign keys, we would be all set. We would simply use these features to implement the various types of keys that are listed in the final DBDL version of the information-level design. Unfortunately, very few systems support all these types of keys. In fact, surprising as it may seem, many of the well-known and widely respected microcomputer DBMSs *don't support any of them*! So we need to devise a scheme for handling these keys in the event that the DBMS cannot do so.

Basically, such a scheme must ensure the uniqueness of primary and candidate keys. It must ensure that values in the foreign keys are legitimate; in other words, that they match the value of the primary key on some row in another table. As far as secondary keys are concerned, we merely need to ensure the efficiency of access to rows on the basis of a value of the secondary key.

For instance, suppose we are implementing the *EMPLOYEE* table shown in Figure 6.1a, in which *EMPNUMB* is the primary key, *SSNUMB* is a candidate key, *EMPNAME* is a secondary key, and *DEPTNUMB* is a foreign key that matches the *DEPT* table. We will have to ensure that the following conditions hold true:

1. Employee numbers are unique.
2. Social security numbers are unique.
3. Access to an employee on the basis of his or her name is rapid. (This restriction differs in that it merely states that a certain type of activity must be efficient, but it is an important restriction nonetheless.)
4. Department numbers are valid; that is, they match the number of a department currently in the database.

The next question is, Who should enforce these restrictions? Two choices are possible, provided the DBMS can't do it. The users of the system could enforce them, or programmers could. In the case of users, they would have to be careful when entering data not to enter two employees with the same employee number, not to enter an employee whose department number was invalid, and so on. Clearly, this would put a tremendous burden on the user.

Provided the DBMS can't enforce the restrictions, the appropriate place for the enforcement to take place is in programs. Thus, the responsibility burden for this enforcement should fall on the programmers who write the programs that users will run to update the database. Incidentally, users *must* update the data through these programs and *not* through the built-in features of the DBMS in such circumstances; otherwise, they would be able to bypass all the controls that we are attempting to program into the system.

Thus, it is the responsibility of programmers to include logic in their programs to enforce all the constraints. With respect to the DBDL shown in Figure 6.1a, this means the following:

1. Before an employee is added, the program should determine three things:
   a. whether an employee with the same employee number is already in the database; if so, the update should be rejected.

      b. whether an employee with the same social security number is already in the database; if so, the update should be rejected; and

      c. whether the department number that was entered matches the number of a department that is already in the database; if it doesn't, the update should be rejected.

2. When an employee is changed, if the department number is one of the values that is changed, the program should check to make sure that the new number also matches the number of a department that is already in the database. If it doesn't, the update should be rejected.

3. When a department is deleted, the program should check to make sure that the database contains no employees for this department. If the department does contain employees and it is allowed to be deleted, these employees will have department numbers that are no longer valid. In that case, the update should be rejected.

These actions must be performed efficiently, and in most systems this means creating and using indexes on all key columns. Thus, an index will be created for each column (or combination of columns) that is a primary key, a candidate key, a secondary key, or a foreign key.

SUMMARY

1. Database design is the process of determining an appropriate database structure to satisfy a given set of requirements. It is a two-part process:
   a. The information-level design, wherein a clean DBMS-independent design is created to satisfy the requirements.
   b. The physical-level design, wherein the final information-level design is converted into an appropriate design for the particular DBMS that will be used.
2. A user view is the view of data necessary to support the operations of a particular user. In order to simplify the design process, the overall set of requirements is split into user views.
3. The information-level design methodology involves applying the following steps to each user view:
   a. Represent the user view as a collection of tables.
   b. Normalize these tables; that is, convert this collection into an equivalent collection that is in 3NF.
   c. Represent all keys: primary, alternate, secondary, and foreign.
   d. Merge the results of the previous step into the cumulative design.
4. The design is represented in a language called DBDL (Database Design Language).
5. To obtain a pictorial representation of a design, apply the following steps to the DBDL design:
   a. Create a rectangle for each table in the DBDL design.
   b. For each foreign key, create an arrow that (1) begins with the rectangle that corresponds to the table identified by the foreign key and (2) terminates at the rectangle that corresponds to the table containing the foreign key. The foreign key may then be removed, although it is not essential to do so.
6. Assuming that a relational or relational-like microcomputer DBMS is going to be used, the physical-level design process consists of creating a table for each table in the DBDL design. Any constraints (primary key, alternate key, or foreign key) that the DBMS cannot enforce must be enforced by the programs in the system, so this fact must be documented for the programmers.

KEY TERMS

alternate key   physical-level design
database design   primary key
Database Design Language   referential integrity
  (DBDL)   relationship
foreign key   secondary key
information-level design   user view
integrity rules

EXERCISES

1. Define the term *user view* as it applies to database design.
2. What is the purpose of breaking down the overall design problem into a consideration of individual user views?
3. Under what circumstances would you not have to break down the overall design into a consideration of individual user views?
4. The information-level design methodology presented in this section contains a number of steps that are to be repeated for each user view. List the steps and briefly describe the kinds of activities that must take place at each step.
5. Describe the function of each of the following types of keys: primary, alternate, secondary, and foreign.
6. Describe the process of mapping an information-level design to a design for a relational model system.
7. Suppose that a given user view contains information about employees and projects. Suppose further that each employee has a unique employee number and that each project has a unique project number. Explain how you would implement the relationship between employees and projects in each of the following scenarios:
   a. Many employees can work on a given project, but each employee can work on only a *single* project.
   b. An employee can work on many projects but each project has a *unique* employee assigned to it.
   c. An employee can work on many projects, and a project can be worked on by many employees.
8. A database at a college is required to support the following requirements:
   a. For a department, store its number and name.
   b. For an advisor, store his or her number and name and the number of the department to which he or she is assigned.
   c. For a course, store its code and description (for example, MTH110, ALGEBRA).
   d. For a student, store his or her number and name. For each course the student has taken, store the course code, the course description, and the grade received. Also, store the number and name of the student's advisor. Assume that an advisor may advise any number of students but that each student has just one advisor.
   Complete the information-level design for this set of requirements. Use your own experience to determine any constraints you need that are not stated in the problem. Represent the answer in DBDL.
9. List the changes that would need to be made in your answer to Exercise 8 if a student could have more than one advisor.

10. Suppose that in addition to the requirements specified in Exercise 8, we must store the number of the department in which the student is majoring. Indicate the changes this would cause in the design in these two situations:
    a. The student must be assigned an advisor who is in the department in which the student is majoring.
    b. The student's advisor does not necessarily have to be in the department in which the student is majoring.
11. Illustrate mapping to the relational model by means of the design shown in Figure 6.13. List the relations. Identify the keys. List the special restrictions that programs will have to enforce.
12. In Example 2 of section 6.4, the claim was made that User View 6 led to the same set of tables that had been created for User View 5. Show that this is true.

Functions of a Database Management System

OBJECTIVES

1. Discuss the following eight functions, or services, that should be provided by a DBMS:
   a. data storage, retrieval, and update
   b. a user-accessible catalog
   c. support for shared update
   d. backup and recovery services
   e. security services
   f. integrity services
   g. services to promote data independence
   h. utility services
2. Discuss the manner in which these services are typically provided.

7.1 INTRODUCTION

A good DBMS should furnish a number of capabilities. As you might expect, the list of features that a full-scale mainframe DBMS could provide would be more extensive than a comparable list for a microcomputer system. Moreover, mainframe systems often furnish these features in a more sophisticated fashion. If you are interested in the features of mainframe systems, see Chapter 2 of [14]. In this chapter, we will focus on microcomputer systems. The list of features that a microcomputer DBMS should furnish includes the following:

1. **Data storage**, **retrieval**, and **update**: the ability to store, retrieve, and update the data that is in the database.
2. A user-accessible **catalog** in which descriptions of data items are stored and which is accessible to users.
3. Support for **shared update**: a mechanism to ensure accuracy when several users are updating the database at the same time.
4. **Backup** and **recovery** services: a mechanism for recovering the database in the event that the database is damaged in any way.
5. **Security** services: a mechanism to ensure that only authorized users can access the database.
6. **Integrity** services: a mechanism to ensure that certain rules are followed with regard to data in the database and any changes that are made in the data.
7. **Services** to promote **data independence**: facilities to support the independence of programs from the structure of the database.
8. **Utility** services: DBMS-provided services that assist in the general maintenance of the database.

The preceding list is summarized in Figure 7.1.

Figure 7.1

Functions of a
DBMS

Functions of a DBMS

1. Data storage, retrieval, and update
2. A user-accessible catalog
3. Support for shared update
4. Backup and recovery services
5. Security services
6. Integrity services
7. Services to promote data independence
8. Utility services

7.2   STORAGE AND RETRIEVAL

A DBMS must furnish users with the ability to store, retrieve, and update the data that is in the database.

This statement about storage and retrieval almost goes without saying. It defines the fundamental capability of a DBMS. Unless a DBMS provides this facility, further discussion of what a DBMS can do is irrelevant. In storing, updating, and retrieving data, it should not be incumbent upon the user to be aware of the system's internal structures or the procedures used to manipulate these structures. This manipulation is strictly the responsibility of the DBMS.

7.3   CATALOG

A DBMS must furnish a **catalog** in which descriptions of data items are stored and which is accessible to users.

This catalog contains crucial information for those who are in charge of a database or who are going to write programs to access a database. Such persons must be able to easily determine what the database "looks like." Specifically, they need to be able to get quick answers to questions like the following:

1. What tables and columns are included in the current structure? What are their names?
2. What are the characteristics of these columns? For example, is the *CUSTNAME* column within the *CUSTOMER* row 20 characters long or 30? Is the *CUSTNUMB* column a numeric field or is it a character field? How many decimal places are in the *UNITPRCE* column?
3. What are the possible values for the various columns? Are there any restrictions on the possibilities for *CREDLIM*, for example?
4. What is the meaning of the various columns? For example, what exactly is *ITEMCLSS*, and what does the item class HW mean?
5. What relationships are present? What is the meaning of each relationship? Must the relationship always exist? For example, must a customer always have a sales rep?
6. Which programs within the system access which data within the database? How do they access it? Do they merely retrieve the data, or do they update it? What kinds of updates do they do? Can a certain program add a new customer, for example, or can it merely make changes regarding information about customers whose names are already in the database? When it makes a change with regard to a customer, can it change all the columns or only the address?

Mainframe DBMSs are often accompanied by a separate entity called a **data dictionary**, which contains answers to all of the previous questions and more. The data dictionary forms a sort of super catalog. Microcomputer DBMSs are not typically accompanied by such a comprehensive tool, but they often have built-in capabilities that furnish answers to at least some of these questions. At a minimum, the capabilities they furnish would allow us to obtain the answers to questions 1 and 2 in the preceding list. Some microcomputer DBMSs provide features that allow us to ask questions along the lines of 3, 4, and 5.

7.4 SHARED UPDATE

A DBMS must furnish a mechanism to ensure accuracy when several users are updating the database at the same time.

Microcomputer databases are often used by just one person at one machine. Sometimes several people may be allowed to update a database but only one person at a time. For example, several people might take turns with one microcomputer to access the database. The advent of microcomputer networks and microcomputer DBMSs that were capable of running on these networks and of allowing several users to access the same database gave rise to a problem that had been a headache to mainframe database management for years: shared update.

By **shared update**, we mean that two or more users are involved in making updates to the database at the same time. On the surface, it might seem that shared update wouldn't present any problem. Why couldn't two or three (or fifty, for that matter) users update the database simultaneously without incurring a problem?

The Problem   To illustrate the problems involved in shared update, let's assume that we have two users, Fred and Sue, who both work for Premiere Products. Fred is currently accessing the database to process orders and, among other things, to increase customers' balances by the amount of the orders. Let's say that Fred is going to increase the balance of customer 124 (Sally Adams) by $100.00. Sue, on the other hand, is accessing the database to post payments and, among other things, to decrease customers' balances by the amount of the payments. As it happens, customer 124 has just made a $100.00 payment, so Sue will decrease her balance by $100.00. The balance of customer 124 was $418.75 prior to the start of Fred and Sue's activity and, since the amount of the increase exactly matches the amount of the decrease, the balance should still be $418.75 after the activity has been completed. But will it? That depends.

How exactly does Fred make the required update? First, the data concerning customer 124 is read from the database into Fred's work area. Second, any changes are made in the data in his work area; in this case, $100.00 is added to the current balance of $418.75, bringing the balance to $518.75. This change has *not* yet taken place in the database, *only* in Fred's work area. Finally, the information is written to the database and the change is now made in the database itself (see Figure 7.2).

Figure 7.2

Fred updates the database

Fred's Work Area   Database on Disk   Sue's Work Area

Prior to Step 1 — Before updates

(continued)

Figure 7.2

(continued)

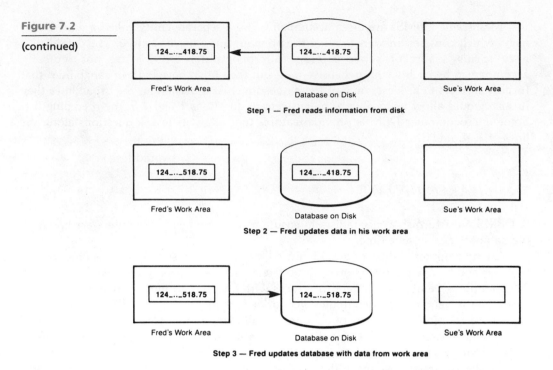

Step 1 — Fred reads information from disk

Step 2 — Fred updates data in his work area

Step 3 — Fred updates database with data from work area

Suppose that Sue begins her update at this point. The data for customer 124 will be read from the database, including the new balance of $518.75. The amount of the payment, $100.00, will then be subtracted from the balance, thus giving a balance of $418.75 *in Sue's work area.* Finally, this new information is written to the database, and the balance for customer 124 is what it should be (see Figure 7.3).

Figure 7.3

Sue updates the database

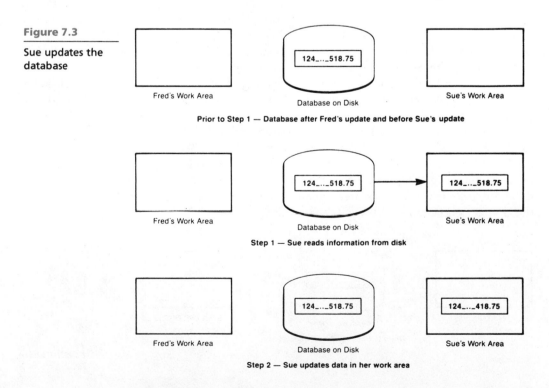

Prior to Step 1 — Database after Fred's update and before Sue's update

Step 1 — Sue reads information from disk

Step 2 — Sue updates data in her work area

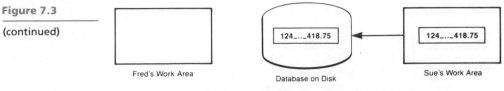

Figure 7.3

(continued)

Fred's Work Area   Database on Disk   Sue's Work Area

Step 3 — Sue updates database with data from work area

In the preceding scenario, things worked out the right way. But they don't necessarily have to. Do you see how things could happen in a way that would lead to an incorrect result? What if the scenario shown in Figure 7.4 had occurred instead? Here, Fred reads the data from the database into his work area, and at about the same time, Sue reads the data from the database into her work area. At this point, both Fred and Sue have the correct data for customer 124, including a balance of $418.75. Fred adds $100.00 to the balance in his work area, and Sue subtracts $100.00 from the balance in her work area. At this point, in Fred's work area the balance reads $518.75, while in Sue's work area it reads $318.75. Fred now writes to the database. At this moment, customer 124 has a balance of $518.75 in the database. Then Sue writes to the database. Now the balance for customer 124 in the database is *$318.75!* (This is a very good deal for Sally Adams, but not such a good deal for Premiere Products.) Had the updates taken place in the reverse order, the final balance would have been $518.75. In either case, we now have incorrect data in our database (one of the updates has been *lost*). This cannot be permitted to happen.

Figure 7.4

Fred and Sue update the database in a manner that leads to inconsistent data

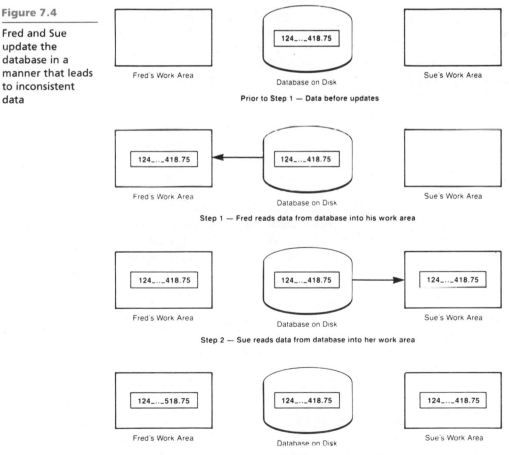

Fred's Work Area   Database on Disk   Sue's Work Area

Prior to Step 1 — Data before updates

Fred's Work Area   Database on Disk   Sue's Work Area

Step 1 — Fred reads data from database into his work area

Fred's Work Area   Database on Disk   Sue's Work Area

Step 2 — Sue reads data from database into her work area

Fred's Work Area   Database on Disk   Sue's Work Area

Step 3 — Fred updates data in his work area

(continued)

Figure 7.4

(continued)

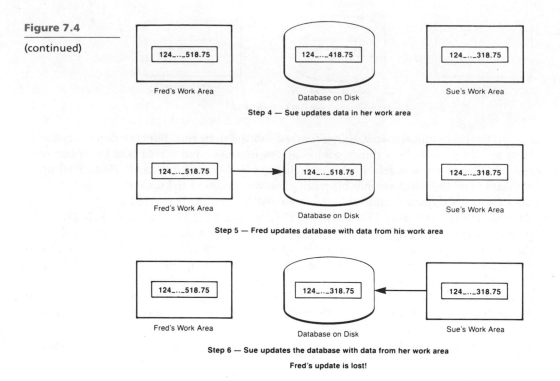

Step 4 — Sue updates data in her work area

Step 5 — Fred updates database with data from his work area

Step 6 — Sue updates the database with data from her work area

Fred's update is lost!

Avoiding the Problem

One way to prevent this situation from occurring is to prohibit shared update. This may seem a little drastic, but it is not really so farfetched. We could permit several users to access the database at the same time, but for *retrieval* only; that is, they would be able to read information *from* the database but they would not be able to write anything *to* the database. When these users entered some kind of transaction to update the database (like posting a payment), the database itself would not be updated at all. Instead, a record would be placed in a separate file of transactions. A record in this file might indicate, for example, that $100.00 had been received from customer 124 on a certain date. Periodically, a single update program would read the records in this transaction file and perform the appropriate updates to the database. Since this program would be the only one to update the database, we would eliminate the problems associated with shared update.

While this approach would avoid one set of problems, it would create another. From the time users started updating, that is, placing records in the update files, until the time the update program actually ran, the data that was in the database would be out of date. Where a customer's balance in the database was $49.50, it would actually be $649.50 if a transaction had been entered that increased it by $600.00. If the customer in question had an $800 credit limit, he or she should be prohibited from charging, say, a $200 item. But according to the data currently in the database, this would not be so. On the contrary, the data in the database would indicate that this customer still had $750.50 of available credit ($800 − $49.50). In a situation that requires the data in the database to be current, this scheme for avoiding the problems of shared update will not work.

Locking
Assuming that we cannot solve the shared update problem by avoiding it, we need a mechanism for dealing with the problem. We need to be able to keep Sue from even beginning the update on customer 124 until Fred has completed his update (or vice versa). This can be accomplished by some kind of **locking** scheme. Suppose that once Fred had read the row for customer 124, it became locked (no other user could access it) and remained locked until Fred had completed the update. For the duration of the lock, any attempt by Sue to read the row would be rejected, and she would be notified that the row was locked. If she chose to do so, she could keep attempting to read the row until it was no longer locked, at which time she could begin her update. This scenario is demonstrated in Figure 7.5. In at least this simple case, the problem of a "lost update" seems to have been solved.

Figure 7.5

Fred and Sue update the database. Locking prevents inconsistent data

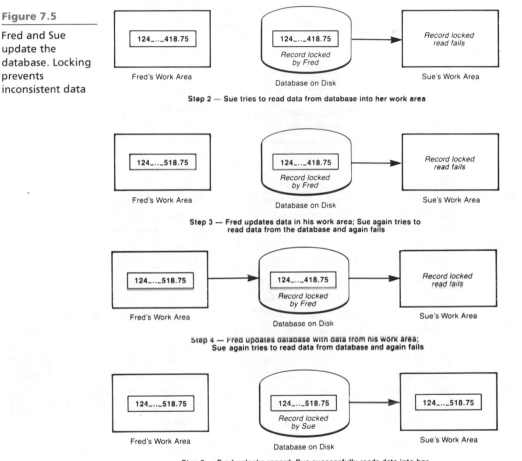

(continued)

Figure 7.5

(continued)

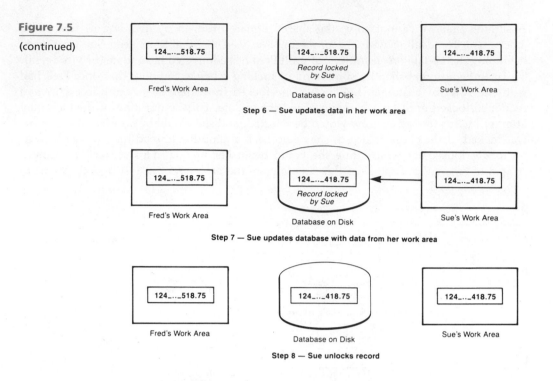

Step 6 — Sue updates data in her work area

Step 7 — Sue updates database with data from her work area

Step 8 — Sue unlocks record

Duration of Locks

How long should a lock be held? If the update involves just changing some values in a single row in a single table (like changing the name of a customer), the lock is no longer necessary once this row has been updated. Sometimes, however, the situation is more involved.

Consider, for example, the process of filling an order. To a user sitting at a terminal, this may seem to involve a single action. A user may merely indicate that an order which is currently on file now needs to be filled. Or he or she may also be required to enter certain data on the order. In either case, however, the process still feels like a single action to the user. Behind the scenes, though, lots of activity may be taking place. We might have to update the *UNONHAND* column in the *PART* table for each part that is on the order in order to reflect the number of units of that part that were shipped and that are consequently no longer on hand. We might also have to update the *BALANCE* column in the *CUSTOMER* table for the customer who placed the order, increasing it by the total amount of the order. Or we might have to update the *TOTCOMM* column in the *SLSREP* table for the sales rep who represents this customer, increasing it by the amount of commission generated by the order.

In circumstances like these, where a single action on the part of a user necessitates several updates in the database, what do we do about locks? How long do we hold each one? For safety's sake, locks should be held until all the required updates have been completed. Serious problems can result from not doing this. (A discussion of such problems is beyond the scope of this text; if you are interested in some of the specifics, see [14].)

Deadlock

Users can hold more than one lock at a time and this may give rise to another potential problem. Let's suppose that Mary has locked the row for customer 124 and is attempting to lock the row for part BT04. Let's also suppose that Tom already has part BT04 locked, so Mary must wait for him to unlock it. Before Tom unlocks part BT04, though, he needs to update, and thus lock, the row for customer 124, which is currently locked by — you guessed it — Mary. Mary is waiting for Tom to act (release the lock for part BT04), while Tom, on the other hand, is waiting for Mary to act (release the lock for customer 124). Without the aid of some outside intervention, this dilemma could be prolonged indefinitely. The term used to describe such situations is **deadlock**, and obviously, some strategy is necessary to either prevent or handle it.

Locking on Microcomputer DBMSs

Mainframe DBMSs typically offer sophisticated schemes for locking as well as for detecting and handling deadlocks (see [14]). Microcomputer DBMSs provide facilities for the same purposes, but they are usually much more limited than the facilities that are provided by mainframe DBMSs. These limitations, in turn, put an additional burden on the programmers who write the programs that allow several users to update the same database simultaneously.

Although the exact features for handling these problems vary from one microcomputer DBMS to another, the following list is fairly typical of the types of facilities provided:

1. Programs can lock a whole table or an individual row within a table (but only one). As long as one program has a row or table locked, no other program may access it.
2. Programs can release any or all of the locks they currently hold.
3. Programs can inquire whether a given row or table is locked.

This list, though it is short, comprises the *complete* set of facilities provided by many systems. Consequently, the following guidelines have been devised for writing programs for a shared-update environment:

1. If more than one row in the same table must be locked during an update, the *whole table* must be locked.
2. When a program attempts to read a row that is locked, it may wait a short period of time and then try to read the row again. This process could continue until the row becomes unlocked. It is usually preferable, however, to impose a limit on the number of times a program may attempt to read the row. In this case, reading is done in a loop, which proceeds until either (a) the read is successful, or (b) the maximum number of times that the program can repeat the operation is reached. Programs vary in terms of what action is taken should the loop be terminated without the read being successful. One possibility is to notify the user of the problem and let him or her decide whether to try the same update again or move on to something else.
3. Since there is no facility to *detect and handle* deadlocks, we must try to *prevent* them. A common approach to this problem is for every program in the system to attempt to lock all of the rows and/or tables it needs before beginning an update. Assuming it is successful in this attempt, each program can then perform the required updates. If any row or table that the program needs is already locked, it should immediately release *all* the locks that it currently holds, wait some specified period of time, and then try the entire process again. (In some cases, it may be better to notify the user of the problem and see whether the user wants to try again.) Effectively, this means that any program that encounters a problem will immediately get out of the way of all the other programs, rather than be involved in a deadlock situation.

4. Since locks prevent other users from accessing a portion of the database, it is important that no user keep rows or tables locked any longer than necessary. This is especially significant for on-line update programs. Suppose, for example, that a user is employing some on-line update program to update customers. Suppose further that once the user enters the number of the customer to be updated, the customer row is locked and remains locked until the user has entered all the new data and the update has taken place. What if the user is interrupted by a phone call before he or she has completed filling in the new data? What if the user goes to lunch? The row might remain locked for an extended period of time. If the update involves several rows, all of which must be locked, the problem becomes that much worse. In fact, in many microcomputer DBMSs, if more than one row from the same table must be locked, the whole table must be locked, which means that whole tables may be locked for extended periods of time. Clearly, this cannot be permitted to occur.

The trick here is for programs to read the information they need at the beginning of the update and then immediately *release all locks*. After the user has entered all the new data, the update takes place as described earlier (that is, attempt to lock all required rows; proceed with the update if successful; release all locks if not successful). This does pose a problem, however. Suppose my program read the data for customer 124 and then released its lock on this customer while the user of my program was filling in new data on the screen. What if a user of your program updated customer 124 in the meantime and the update was completed before my user finished filling in the new data? If my user then were to finish filling in the new data and my program blindly went ahead to update the row for customer 124 with this new data, your user's update would be *lost* (that is, over-written with my user's data). So, my program should take a further precautionary step. Before blindly updating the database with my user's data, my program should make sure that nobody else has updated the data in the meantime. If someone has, my program cannot update the database with my user's data; instead, my user must be informed of the situation and permitted to decide whether he or she wants to redo the update or move on to something else.

How will my program know whether or not some other user has updated the row for customer 124? Several methods can be used to provide the answer. One is to include an additional column in each row, perhaps a three-digit number called *UPDCOUNT*. Every time any program updates a row in any way, it should also update the value in this column by adding 1 to it. (If the previous value was 999, the new value produced by adding one to the old value would be too big for the column. In such a case, the program updating the row would have to set the update count back to zero.) Assuming that every program in the system were to adhere to this approach, we could utilize the following logic:

1. Read all the data from the row for customer 124, including the value of *UPDCOUNT*. (Let's assume for the purposes of this example that the value is 478.) Store this value in some variable for future reference. Unlock the row.
2. Get all the new data from the user.
3. When it is time to do the update, lock the row for customer 124, read the current data, and examine the value of *UPDCOUNT*. If it is still the same (in our case, 478), the row has not been updated and we can finish our update. If it is different (479 or 480, for example), we know that at least one other program has updated the data in the meantime and we cannot complete our update.
4. If we have to lock multiple rows, the same procedure is followed for each one; that is, for each of the rows involved, store its update count in some variable. When it is time to do the update, lock all the rows, read each row's update count, and compare it with the count we have stored. If the counts *all* agree, we can perform the update. If they don't agree, the update cannot take place.

Two crucial points arise from the preceding discussion. First, the logic to support shared update certainly adds a fair amount of complexity to each of the programs in the system. Second, cooperation among programs is *essential*. Every program must do its job. If one program doesn't update the *UPDCOUNT* column, for example, another program may assume that a row has not been updated when, in fact, it has been. If a program doesn't release all its locks when it encounters a row or table it needs that is locked by some other program, the possibility of deadlock arises. If a program does not release its locks while its user is entering data on the screen, the performance of the whole system may suffer.

One might naturally ask at this point, Is it worth it? Is the ability to have several users updating the database simultaneously worth the complexity that it adds to every program in the system? In some cases, the answer will be no. Shared update may be far from a necessity. In other cases, however, shared update will be necessary to the productivity of the users of the system. In these cases, implementation either of the ideas we have been discussing or of some similar scheme is essential to the proper performance of the system.

7.5   RECOVERY

A DBMS must furnish a mechanism for recovering the database in the event that the database is damaged in any way.

A database can be damaged or destroyed in a number of ways. Users can enter data that is incorrect; programs that are updating the database can abort during an update; a hardware problem can occur; and so on. After such an event has occurred, the database may contain invalid data. It may even have been totally destroyed.

Obviously, a situation in which data has been damaged or destroyed cannot be allowed to go uncorrected. The database must be returned to a correct state. This process is called **recovery**; we say that we **recover** the database.

The simplest approach to recovery involves periodically making a copy of the database (called a **backup**, or a **save**). If a problem occurs, the database is recovered by copying this backup copy over it. In effect, the damage is undone by returning the database to the state it was in when the last backup was made.

Unfortunately, other activity besides that which caused the destruction is also undone. Suppose the database is backed up at 10:00 at night and users begin updating it at 8:00 the next morning. Suppose further that at 11:30 that morning, something happens that destroys the database. If the previous night's backup is used to recover the database, the *entire* database is returned to the state it was in at 10:00 the previous night. *All* updates made in the morning are lost, not just the update or updates that were in progress at the time the problem occurred. This would mean that during the final part of the recovery process, users would have to redo all the work they had done since 8:00 in the morning.

As you might expect, mainframe DBMSs provide sophisticated facilities to avoid the costly and time-consuming process of having users redo their work. These facilities maintain a record (called a **journal**) of all updates to the database. (If you are interested in the manner in which these features work, see Chapter 2 of [14].)

Such features are not generally available at this time on microcomputer DBMSs. Most of them furnish users with a simple way to make backup copies and to recover the database later by copying the backup over the database; but this is all they furnish in this regard.

Given this state of affairs, how should we handle backup and recovery in any application system we develop with a microcomputer DBMS? We could simply use the features of the DBMS to periodically make backup copies, and use the most recent backup if a recovery were necessary. The more crucial it was to avoid redoing work, the more often we would make backup copies. (If a backup were made every eight hours, for example, we might have to redo up to eight hours of work. If one were made every two hours, on the other hand, at most two hours of work would have to be redone.)

In many situations, this approach, although not particularly desirable, is acceptable. For some systems, however, it is not. In such cases, the necessary recovery features that are not supplied by the DBMS must be included in the application programs. Each of the programs that update the database could, for example, also write a record to a separate file, the journal, indicating the update that had taken place. A separate program could be written that would look at this file and recreate all of the updates indicated by the records in the file. The recovery process would then consist of (1) copying the backup over the actual database and (2) running this special program.

While this approach does simplify the recovery process for the users of the system, it also causes some problems. First, each of the programs in the system becomes more complicated because of the extra logic involved in adding records to the special file. Second, a separate program to update the database with the information in this file must be written. Finally, every time a user completes an update, the system now has extra work to do, and this additional processing may slow down the system to an unacceptable pace. Thus, in any application, we must determine whether the ease of recovery furnished by this approach is worth the price we may have to pay for it. The answer will vary from one system to another.

7.6 SECURITY

A DBMS must furnish a mechanism that restricts access to the database to authorized users. The term **security** refers to the protection of the database against unauthorized (or even illegal) access, either intentional or accidental. The most common features used by microcomputer DBMSs to provide for security are passwords and encryption.

Passwords

Many microcomputer DBMSs furnish sophisticated schemes whereby system administrators can assign passwords. Each password may be associated with a list of actions that the user who furnishes it is permitted to take. A user who furnished the password XY1JE, for example, might be allowed to view and alter any customer data, whereas another user who furnished the password GS36Y might be permitted to view and alter a customer's name or address, view but not alter a customer's credit limit, and not even view a customer's balance.

Encryption

Encryption refers to the storing of the data in the database in an encrypted format. Any time a user stores or modifies data in the database, the DBMS will encrypt the data before actually updating the database. Before a legitimate user retrieves the data via the DBMS, the data will be decrypted. The whole encryption process is transparent to a legitimate user; that is, he or she is not even aware it is happening. However, if an unauthorized user attempts to bypass all the controls of the DBMS and get to the database directly, he or she will be able to see only the encrypted version of the data.

Views   If a DBMS provides a facility that allows various users to have their own **views** of a database, this can be used for security purposes. Tables or columns to which the user does not have access in his or her view effectively do not exist for that user.

For further discussion of security, including security features found on mainframe systems, see [14].

7.7 INTEGRITY

A DBMS must furnish a mechanism to ensure that both the data in the database and changes in the data follow certain rules.

In any database, there will be conditions, called **integrity constraints**, that must be satisfied by the data within the database. The types of constraints that may be present fall into the following four categories:

1. **Data type.** The data entered for any column should be consistent with the data type for that column. For a numeric column, only numbers should be allowed to be entered. If the column is a date, only a legitimate date (in the form MMDDYY or MM/DD/YY) should be permitted. For instance, 13/07/92 would be an illegitimate date that should be rejected by the DBMS.

2. **Legal values.** It may be that for certain columns, not every possible value that is of the right type is legitimate. For example, even though *CREDLIM* is a numeric column, only the values 300, 500, 800, and 1,000 may be valid. It may be that only numbers between 2.00 and 800.00 are legal values for *UNITPRCE*.

3. **Format.** It may be that certain columns have a very special format that must be followed. Even though the column *PARTNUMB* is a character field, for example, only specially formatted strings of characters may be acceptable. Legitimate part numbers may have to consist of two letters followed by a hyphen, followed by a three-digit number. This is an example of a format constraint.

4. **Key constraints.** There are two types of key constraints: primary key constraints and foreign key constraints. Primary key constraints enforce the uniqueness of the primary key. For example, forbidding the addition of a sales rep whose number matched the number of a sales rep already in the database would be a primary key constraint. Foreign key constraints enforce the fact that a value for a foreign key must match the value of the primary key for some row in another table. Forbidding the addition of a customer whose sales rep *was not already in the database* would be an example of a foreign key constraint.

An integrity constraint can be treated in one of four ways:

1. The constraint can be ignored, in which case no attempt is made to enforce the constraint.
2. The burden of enforcing the constraint can be placed on the users of the system. This means that users must be careful that any changes they make in the database do not violate the constraint.
3. The burden can be placed on programmers. Logic to enforce the constraint is then built into programs. Users must update the database only by means of these programs and not through any of the built-in entry facilities provided by the DBMS, since these would allow violation of the constraint. The programs are designed to reject any attempt on the part of the user to update the database in such a way that the constraint is violated.
4. The burden can be placed on the DBMS. The constraint is specified to the DBMS, which then rejects any attempt to update the database in such a way that the constraint is violated.

Q & A

Question:	Which of these approaches is best?
Answer:	The fourth approach. Here is why.

The first approach is undesirable, since it can lead to invalid data in the database (two customers with the same number, part numbers with an invalid format, illegal credit limits, and so on).

The second approach is a little better, since at least an attempt is made to enforce the constraints. Yet it puts the burden of enforcement on the user. Not only does this mean extra work for the user, but any mistake on the part of a single user, no matter how innocent, can lead to invalid data in the database.

The third approach removes the burden of enforcement from the user and places it on the programmers. This is better still, since it means that users will be unable to violate the constraints. The disadvantage is that all of the update programs in the system are made more complex. This complexity makes the programmers less productive and makes the programs more difficult to create and to modify. It also makes changing an integrity constraint more difficult, since this may mean changing all of the programs that update the database. Further, any program in which the logic that is used to enforce the constraints is faulty could permit some constraint to be violated *without our even being aware that this had happened* until some problem that occurred at a later date brought it to our attention. Finally, we would have to carefully guard against a user bypassing the programs in the system in order to enter data directly into the database (for example, by using some built-in facility of the DBMS). If this should happen, all of the controls we had so diligently placed into our programs would be helpless to prevent a violation of the constraints.

The best approach is the one in which we put the burden on the DBMS. We would specify any constraints to the DBMS and the DBMS would ensure that they are never violated.

Unfortunately, most microcomputer DBMSs don't have all the necessary capabilities to enforce the various types of integrity constraints (neither do many mainframe DBMSs). Usually, the approach that is taken is a combination of the (3) and (4) in the foregoing list. We let the DBMS enforce any of the constraints that it is capable of enforcing; other constraints are enforced by application programs. We might also create a special program whose sole purpose would be to examine the data in the database to determine whether any constraints had been violated; this program would be run periodically. Corrective action could be taken to remedy any violations that were discovered by means of this program.

Current Microcomputer DBMSs and Integrity

We conclude this section with a discussion of the constraints that current microcomputer DBMSs are able to enforce.

- Virtually all microcomputer DBMSs do an excellent job of enforcing data-type constraints. At a minimum, they typically allow data types of numeric, character, and date. Users are prevented from entering nonnumeric data into numeric columns or invalid dates into date columns.

- Most microcomputer DBMSs do not provide direct support for enforcing constraints that involve legal values. In cases where they do provide such support, it is often for a range of numbers (unit price must be between 2.00 and 8.00), but not for selected numbers (credit limit must be 300, 500, 800, or 1,000). Of the systems that don't provide direct support, some will supply support if users update the data through custom-generated forms; in other words, the constraints can be specified during the description of a form and the constraints will be enforced for any user who employs that form to update the data in the database. However, if the database is updated in some other way, these constraints will not be enforced.

■ The way in which many current microcomputer DBMSs treat format constraints is similar to the way in which they treat legal values. The constraints are not implemented directly, but some fairly sophisticated format-type constraints can be built into custom-created forms. In the case of some DBMSs, however, a substantial number of format constraints can be communicated to the DBMS directly, and these constraints will be enforced no matter how the data is entered into the database.

■ Most microcomputer DBMSs (and many mainframe DBMSs) are weakest in the area of key constraints. Many DBMSs do not allow a primary key to be specified and thus will certainly not enforce any uniqueness. Most systems allow users to build an index on the column or columns that constitute the primary key, but many systems do not allow users to specify the uniqueness that is essential for the primary key. Thus, the burden of enforcing the uniqueness will typically be placed on the programs. True support for foreign key constraints is almost totally lacking and the burden of enforcing these constraints (like the requirement that the sales rep for a given customer already be in the database) will also be placed on programs.

With time, the better DBMSs will improve in this area. And as each improvement takes place, some of the burden that was formerly placed on programs will be placed where it belongs — on the DBMS.

For a further discussion of integrity, including the integrity features found on mainframe systems, see [14].

7.8   DATA INDEPENDENCE

A DBMS must include facilities that provide programs with independence in terms of their relationship to the structure of the database.

Some of you may have written or worked with application systems that accessed a collection of files. Were any changes ever required in the types of data stored in the files? Did users ever propose any further requirements that necessitated the addition of columns? What about changing the characteristics of a column (for example, expanding the number of characters in the *NAME* column from 25 to 30)? What about additional processing requirements (for example, a new requirement to rapidly access a customer on the basis of his or her name)?

If any of these things have happened to you, you know that even the simplest of them can be very painful. Adding a new column or changing the characteristics of an existing column, for example, usually entails writing a program that will read each record from the existing file and will write a corresponding record with the new layout to a new file. In addition, each of the programs in the existing system must be changed to reflect the new layout, and these changes must be tested.

One of the advantages of working with a DBMS is **data independence**, that is, the property that changes can be made in the layout of a database without application programs necessarily being affected. Let's examine how the various types of changes that can be made in the structure of the database would affect programs that access the database, we are using a good microcomputer DBMS.

1. **Addition of a column.** No program *should* need to be changed except, of course, those programs which will utilize the new column. Some programs *may* need to be changed, however. If, for example, a program used something like the SQL "SELECT * FROM ..." to select all of the columns from a given table, the user would suddenly be presented with an extra column. To prevent this from happening, the output of the program would have to be restricted to only the desired columns. To avoid the imposition of this extra work, it's a good idea to list specific columns in a SQL SELECT command instead of using the *.

2. **Changing the length of a column.** In general, programs should not have to change because the length of a column has been changed. For the most part, the DBMS will handle all the details concerning this change in length. If, however, a program is designed to set aside a certain portion of the screen or a report for the column and the length of the column has increased to the point where the previously allocated space is inadequate, clearly the program will need to be changed.

3. **Creating a new index.** Typically, a simple command is all that is required to create a new index. But in order to *use* the index, reference must be made to it when the database is opened. Thus, in any program that will use this new index, the statement that opens the database must be changed. In most systems, any index that has been referenced in this way is updated automatically when the data is updated, so no changes are required in those portions of the program which are devoted to updating. To ensure the use of the index in a given query, some special action *may* be required, but the necessary changes are usually very simple ones.

4. **Adding or changing a relationship.** This change is the trickiest of all and is best illustrated with an example. Let's suppose that at Premiere Products we now have the following requirements:

a. Customers are assigned to territories.
b. Each territory is assigned to a single sales rep.
c. A sales rep can have more than one territory.
d. A customer is represented by the sales rep who covers the territory to which the customer is assigned.

To implement these changes, we might choose to restructure the database as follows:

```
SLSREP (SLSRNUMB, SLSRNAME, SLSRADDR,
          TOTCOMM, COMMRATE)
TERRITORY (TERRNUMB, TERRDESC,
          SLSRNUMB)
CUSTOMER (CUSTNUMB, NAME, ADDRESS, BALANCE,
          CREDLIM, TERRNUMB)
```

Now let's suppose that a user is accessing the database via the following view, called *SLSCUST*:

```
CREATE VIEW SLSCUST (SNUMB, SNAME, CNUMB, CNAME) AS
    SELECT SLSREP.SLSRNUMB, SLSREP.SLSRNAME,
         CUSTOMER.CUSTNUMB, CUSTOMER.NAME
         FROM SLSREP, CUSTOMER
         WHERE SLSREP.SLSRNUMB =
              CUSTOMER.SLSRNUMB
```

The defining query is no longer legitimate, since there is no *SLSRNUMB* column in the *CUSTOMER* table. A relationship still exists between sales reps and customers, however. The difference is that we now must go through the *TERRITORY* table to relate the two. If users have been accessing the tables directly to form the relationship, their programs will have to change. If they are using the *SLSCUST* view, then only the definition of the view will have to change. The new definition will be as follows on the next page.

```
CREATE VIEW SLSCUST (SNUMB, SNAME, CNUMB, CNAME) AS
   SELECT SLSREP.SLSRNUMB, SLSREP.SLSRNAME,
      CUSTOMER.CUSTNUMB, CUSTOMER.NAME
   FROM SLSREP, TERRITORY, CUSTOMER
   WHERE SLSREP.SLSRNUMB =
      TERRITORY.SLSRNUMB
      AND TERRITORY.TERRNUMB =
      CUSTOMER.TERRNUMB
```

The defining query is now more complicated than it was before, but this will not affect users of the view. They will continue to access the database in exactly the same way they did before, and their programs will not need to change.

We've now seen how the use of views can allow changes to be made in the logical structure of the database without application programs being affected. As helpful as this is, however, all is not quite as positive as it might seem. For one thing, this entire discussion would not even be relevant to the many DBMSs that do not permit the use of views. Second, even those DBMSs which support views often limit the types of update that can be accomplished through a view. In particular, if the view involves a join, often little or no updating is to be allowed to take place. So the benefits that can be derived from the use of views may very well be unavailable to the user who needs to update the database. This problem is the focus of a great deal of current research and should be resolved in the near future.

7.9   UTILITIES

A DBMS should provide a set of utility services.

In addition to the services already discussed, a DBMS can provide a number of utility-type services that assist in the general maintenance of the database. Following is a list of such services that may be provided by a microcomputer DBMS.

1. Services that permit changes to be made in the database structure (adding new tables or columns, deleting existing tables or columns, changing the name or characteristics of a column, and so on).
2. Services that permit the addition of new indexes and the deletion of indexes that are no longer wanted.
3. Access to DOS services from within the DBMS.
4. Services that provide export to and import from other microcomputer software products. For example, these services allow data to be transferred in a relatively easy fashion between the DBMS and a spreadsheet, word processing, or graphics program.
5. Several of the services that form a part of the fourth-generation environment (see Chapter 9) are also furnished by some of the better microcomputer DBMSs. These include such things as easy-to-use edit and query capabilities, screen generators, report generators, and so on.
6. Access to both procedural and nonprocedural languages. (With a procedural language, the computer must be told precisely how a given task is to be accomplished; BASIC, Pascal, and COBOL are examples of procedural languages. With a nonprocedural language, the task is merely described to the computer, which then determines how to accomplish it. SQL is an example of a nonprocedural language.)
7. An easy-to-use menu-driven interface that allows users to tap the power of the DBMS without having to resort to a complicated set of commands.

SUMMARY

1. A DBMS must furnish users with the ability to store, retrieve, and update data that is in the database.
2. A DBMS must furnish a catalog in which descriptions of the structure of a database are stored and which can be queried by users.
3. A DBMS must provide support for shared update (more than one user updating the database at the same time).
   a. If care is not taken, incorrect results can be produced in the database.
   b. Locking is one approach that ensures correct results. As long as a portion of the database is locked by one user, other users cannot gain access to it.
   c. Deadlock is the term used to describe the situation wherein two or more users are each waiting on the other to give up a lock before they can proceed. Mainframe DBMSs have sophisticated facilities for detecting and handling deadlock. Most microcomputer DBMSs do not have such facilities, which means that programs that access the database must be written in such a way that deadlocks are avoided.
4. A DBMS must provide facilities for recovering the database in the event that it is damaged or destroyed. Most microcomputer DBMSs provide facilities for periodically making a backup copy of the database. To recover the database when it is damaged or destroyed, the backup is copied over the database.
5. A DBMS must provide security facilities; that is, features that prevent unauthorized access to the database. Such facilities typically include passwords; encryption (the storing of data in an encoded form); and views (which limit users to accessing only the tables and columns included in the view).
6. An integrity constraint is a rule that data in the database must follow. A DBMS should include features that prevent integrity constraints from being violated.
7. A DBMS must include facilities that promote data independence, the property that the database structure can change without application programs necessarily being affected.
8. The DBMS must provide a set of utility services.

KEY TERMS

backup	integrity	save
catalog	journal	security
data dictionary	locking	shared update
data independence	lost update	utility services
deadlock	passwords	view
encryption	recovery	

EXERCISES

1. What do we mean when we say that a DBMS should provide facilities for storage, retrieval, and update?
2. What is the purpose of the catalog? What types of information are usually found in the catalogs that accompany microcomputer DBMSs? What additional types of information are often found in the catalogs that accompany mainframe DBMSs?
3. What is meant by shared update?
4. Describe a situation, other than the one given in the text, in which uncontrolled shared update would produce incorrect results.
5. What is meant by locking?
6. How long should locks be held?

7. What is deadlock? How does it occur?

8. Are most microcomputer systems capable of detecting and breaking deadlocks?

9. Assuming that we are using a microcomputer DBMS that provides the locking facilities described in the text, how should programs be written to (a) avoid deadlock, (b) guarantee correct results, and (c) keep any individual user from tying up portions of the database for extended periods of time?

10. What is meant by recovery? What facilities are typically provided by microcomputer DBMSs to handle backup and recovery? What main feature is lacking in such facilities? What problems can this cause for users?

11. What is meant by security?

12. How are passwords used by microcomputer DBMSs to promote security?

13. What is encryption? How does it relate to security?

14. How do views relate to security?

15. What is meant by integrity? What is an integrity constraint? Describe four different ways of handling integrity constraints. Which approach is the most desirable?

16. What is meant by data independence? What benefit does it provide?

17. Name some utility services that a DBMS should provide.

8

Database Administration

OBJECTIVES

1. Discuss the need for database administration.
2. Explain the role of DBA in formulating and implementing database policies.
3. Discuss the role of DBA with regard to the data dictionary, user training, and selection and support of DBMS.
4. Discuss the role of DBA in the database design process.

8.1 INTRODUCTION

We have already seen that the database approach confers many benefits. On the other hand, it involves potential hazards, especially when the database serves more than one user. Problems are associated with shared update, as with security: Who is allowed to access various parts of the database, and in what way? How do we prevent unauthorized accesses? Just managing the database involves fundamental difficulties. Each user must be made aware of the database structure, or at least that portion of the database which he or she is allowed to access. Any changes that are made in the structure must be communicated to all users, along with information about how the changes will affect them. Backup and recovery must be carefully coordinated, much more so than in a single-user environment, and this presents another complication.

In order to surmount these problems, the services of a person or group commonly referred to as **database administration (DBA)** are essential. DBA (usually a group rather than an individual) is responsible for supervising both the database and the use of the DBMS. (Sometimes the term DBA is used to refer to the database administrator, the individual who is in charge of this group. Usually the context makes clear which meaning is intended.)

In this chapter, we will investigate the role of DBA, which is summarized in Figure 8.1. In section 8.2, we'll discuss DBA's role in formulating and implementing important policies with respect to the database and its use. In section 8.3, we will examine DBA's role in the use of the data dictionary. In section 8.4, we'll see that DBA plays a crucial role in training various users. We'll discuss DBA's role in the selection and support of the DBMS in section 8.5. Finally, in section 8.6, we'll discuss the role DBA plays in the database design process.

Figure 8.1

Responsibilities of DBA

Responsibilities of DBA

1. Policy Formulation and Implementation
   a. Access Privileges
   b. Security
   c. Planning for Disaster
   d. Archives
2. Data Dictionary Management
3. Training
4. DBMS Support
   a. DBMS Evaluation and Selection
   b. Responsibility for DBMS
5. Database Design

In this text, we'll be focusing on the role of DBA in a microcomputer environment. The role of DBA in a mainframe environment is similar; for a detailed discussion, see [14]. For another perspective on DBA in a microcomputer environment, see [8] and [12].

8.2 POLICY FORMULATION AND IMPLEMENTATION

DBA formulates database policies and communicates these policies to users. DBA is also charged with the implementation of these policies.

Access Privileges

Access to every table and column in the database is not a necessity for every user. Sam, for example, is an employee at Premiere Products whose main responsibility is the inventory. While he may very well need access to the entire *PART* table, does he also need access to the *SLSREP* table? He should probably be able to print inventory reports, but should he be able to change the layout of these reports? Betty, whose responsibility is customer mailings, clearly requires access to customers' names and addresses, but what about their balances or credit limits? Should she be able to change an address? While sales rep 3 (Mary Jones) should be able to obtain information about her own customers, should she be able to obtain the same information about other customers? Figure 8.2 illustrates permitted and denied access for these employees.

We don't have enough information about the policies of Premiere Products to answer the foregoing questions. The DBA, however, must answer questions like these and take steps to ensure that users access the database only in ways to which they are entitled. Policies concerning such access should be clearly documented for and communicated to all concerned parties.

Figure 8.2a

Permitted and denied access for Sam

SLSREP

SLSRNUMB	SLSRNAME	SLSRADDR	TOTCOMM	COMMRATE
3	MARY JONES	123 MAIN,GRANT,MI	2150.00	.05
6	WILLIAM SMITH	102 RAYMOND,ADA,MI	4912.50	.07
12	SAM BROWN	419 HARPER,LANSING,MI	2150.00	.05

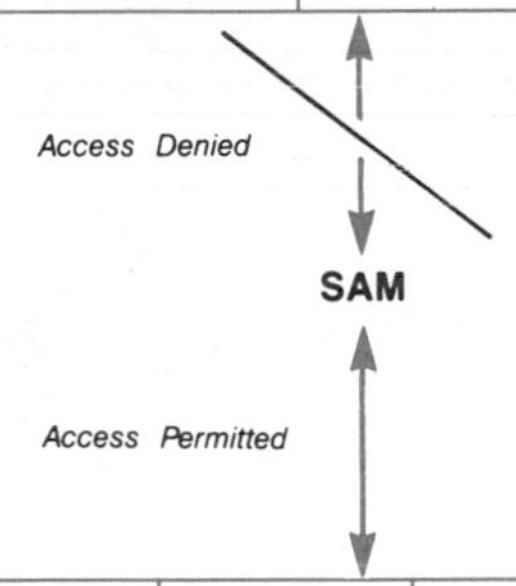

Access Denied

SAM

Access Permitted

PART

PARTNUMB	PARTDESC	UNONHAND	ITEMCLSS	WREHSENM	UNITPRCE
AX12	IRON	104	HW	3	17.95
AZ52	SKATES	20	SG	2	24.95
BA74	BASEBALL	40	SG	1	4.95
BH22	TOASTER	95	HW	3	34.95
BT04	STOVE	11	AP	2	402.99
BZ66	WASHER	52	AP	3	311.95
CA14	SKILLET	2	HW	3	19.95
CB03	BIKE	44	SG	1	187.50
CX11	MIXER	112	HW	3	57.95
CZ81	WEIGHTS	208	SG	2	108.99

Figure 8.2b

Permitted and denied access for Betty

Betty

CUSTOMER

CUSTNUMB	CUSTNAME	ADDRESS	BALANCE	CREDLIM	SLSRNUMB
124	SALLY ADAMS	481 OAK,LANSING,MI	418.75	500	3
256	ANN SAMUELS	215 PETE,GRANT,MI	10.75	800	6
311	DON CHARLES	48 COLLEGE,IRA,MI	200.10	300	12
315	TOM DANIELS	914 CHERRY,KENT,MI	320.75	300	6
405	AL WILLIAMS	519 WATSON,GRANT,MI	201.75	800	12
412	SALLY ADAMS	16 ELM,LANSING,MI	908.75	1000	3
522	MARY NELSON	108 PINE,ADA,MI	49.50	800	12
567	JOE BAKER	808 RIDGE,HARPER,MI	201.20	300	6
587	JUDY ROBERTS	512 PINE,ADA,MI	57.75	500	6
622	DAN MARTIN	419 CHIP,GRANT,MI	575.50	500	3

Access Permitted · *Access Denied*

Figure 8.2c

Permitted and denied access for Mary Jones

CUSTOMER

CUSTNUMB	CUSTNAME	BALANCE	CREDLIM	SLSRNUMB
124	SALLY ADAMS	418.75	500	3
256	ANN SAMUELS	10.75	800	6
311	DON CHARLES	200.10	300	12
315	TOM DANIELS	320.75	300	6
405	AL WILLIAMS	201.75	800	12
412	SALLY ADAMS	908.75	1000	3
522	MARY NELSON	49.50	800	12
567	JOE BAKER	201.20	300	6
587	JUDY ROBERTS	57.75	500	6
622	DAN MARTIN	575.50	500	3

Denied · **Mary Jones** *Permitted Accesses* · *Denied*

Security

As we noted earlier, the term **security** refers to the prevention of unauthorized access to the database. Of course, this includes access by someone who has no right to access the database *at all*, (for example, someone who is not connected with Premiere Products). It can also include users who have legitimate access to some portion of the database but who are attempting to access a portion of it to which they are *not* entitled. Figure 8.3 illustrates both types of security violation.

Figure 8.3a

Attempted security violation: John is not an authorized user

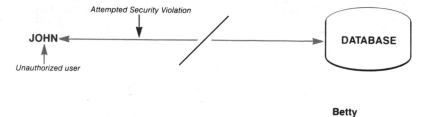

Attempted Security Violation

JOHN ← *Unauthorized user* → DATABASE

Figure 8.3b

Attempted security violation: although Betty is an authorized user, she is not authorized to access customers' balances

Betty

Attempted Security Violation

CUSTOMER

CUSTNUMB	CUSTNAME	ADDRESS	BALANCE	CREDLIM	SLSRNUMB
124	SALLY ADAMS	481 OAK,LANSING,MI	418.75	500	3
256	ANN SAMUELS	215 PETE,GRANT,MI	10.75	800	6
311	DON CHARLES	48 COLLEGE,IRA,MI	200.10	300	12
315	TOM DANIELS	914 CHERRY,KENT,MI	320.75	300	6
405	AL WILLIAMS	519 WATSON,GRANT,MI	201.75	800	12
412	SALLY ADAMS	16 ELM,LANSING,MI	908.75	1000	3
522	MARY NELSON	108 PINE,ADA,MI	49.50	800	12
567	JOE BAKER	808 RIDGE,HARPER,MI	201.20	300	6
587	JUDY ROBERTS	512 PINE,ADA,MI	57.75	500	6
622	DAN MARTIN	419 CHIP,GRANT,MI	575.50	500	3

DBA must take steps to ensure that the database is secure. Once access privileges have been specified, DBA should draw up policies to explain them and should then distribute these policies to authorized users.

Whatever facilities are present in the DBMS, such as passwords, encryption, and/or views, should be utilized by DBA to implement these policies. Any necessary features that the DBMS lacks should be supplemented by DBA through the use of special programs. Figure 8.4 shows security features of the DBMS both with and without DBA.

Figure 8.4a

Security features of DBMS as sole security

Figure 8.4b

Security features of DBMS supplemented by DBA

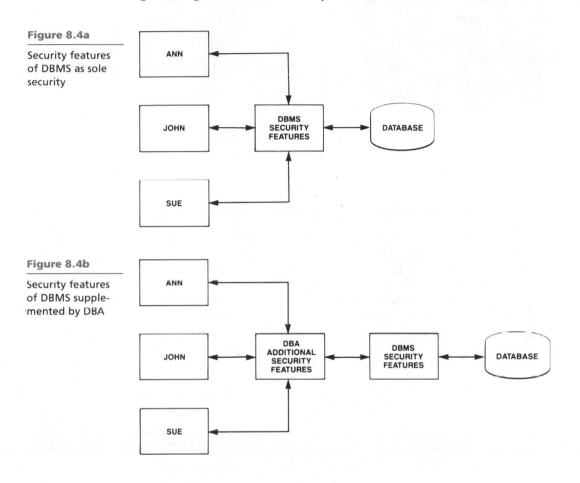

One security feature that we have mentioned, passwords, deserves further attention. Sometimes people think that simply establishing a password scheme will ensure security. After all, Tim can't get access to Pam's data if he doesn't know her password (assuming, of course, that Tim doesn't have a password of his own that allows access to the same data).

But what if Tim observes Pam entering her password? What if he guesses her password? You might think this sort of occurrence is so unlikely that there is nothing to worry about. In fact, it is not so unlikely. Many people often choose passwords that they can remember easily. A very common choice, for example, is the name of a family member. So if Pam is a typical user, Tim might very well be able to obtain her password just by trying names of family members. Other users choose unusual passwords (or have such passwords assigned to them), but, in order to remember them, they often have these passwords written down somewhere. Without giving it much thought, such users may be careless about the paper on which a password is written, giving people like Tim still another vehicle for obtaining one. Figure 8.5 illustrates the careless use of passwords.

Figure 8.5

Careless use of passwords. Tim is trying to look over Pam's shoulder to see the paper on which her password is written

It is up to DBA to educate users on the use of passwords. The pitfalls we have just discussed should be stressed, as should precautionary measures, including the need for frequent changes of passwords.

Other Threats

The type of security we've been discussing concerns harm done by unauthorized users. A database can be harmed in another way as well, and that is through some physical occurrence such as an aborted program, a disk problem, a power outage, a computer malfunction, and so on. This issue was discussed in Chapter 7 in the material on recovery, but it is listed here as well since it is DBA's responsibility to establish and implement backup and recovery procedures. As in other cases, DBA will use the built-in features of the DBMS where possible and will supplement them where they are lacking.

For example, many microcomputer DBMSs lack facilities to maintain a journal (or log) of changes in the database. Thus, recovery is usually limited to copying the most recent backup over the live database. This means, as we have already seen, that any changes made since this backup have to be redone by the users. If this presents a major problem, DBA may decide to supplement the DBMS facilities. A typical solution would be to have each program that updates the database also make appropriate entries in a journal (see Figure 8.6a) and then make use of this journal in the recovery process (see Figure 8.6b). The database is first recovered by copying the backup version over the live database; and then it is brought up-to-date through a special DBA-created program that updates the database with changes recorded in the journal.

Figure 8.6a

Programs involved in database processing also maintain a journal

Figure 8.6b

Journal is used in recovery

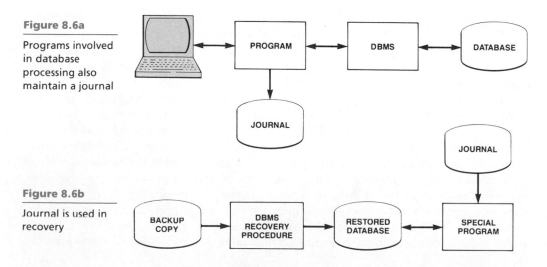

Archives   Often, data needs to be kept in the database for only a limited time. An order that has been filled, has appeared on some statement, and has been paid is in one sense no longer important. Should the order be left in the database? If data is always left in the database as a matter of policy, the database will continually grow. Along with this, the disk space that is occupied by the database expands, and the performance of programs that access the database may deteriorate. Both of these events can ultimately lead to difficulties. This is a reason for removing an already filled order and all of its associated order lines from the database.

On the other hand, it may be necessary to retain such data for future reference. Possible reasons could include customer inquiries, government regulations, auditing requirements, and so on. Apparently, we may have a conflict on our hands: we may need to remove something, and yet we may not be able to afford to remove it.

The solution is to use what is known as a **data archive**. In ordinary usage an archive (technically archives) is a place where public records and documents are kept. A data archive is similar. It is a place where a record of certain corporate data is kept. In fact, we often refer to a data archive simply as an archive. In the case of the aforementioned order, we would remove it from the database and place it in the archive, thus storing it for future reference (see Figure 8.7).

Typically, the archive will be kept on some magnetic form (for example, a disk, a diskette, or a tape). On mainframes, the most common medium for an archive is tape. On micros, tape might also be used, but another common choice is a collection of diskettes. Still another option would be to keep the data in printed form only. In any case, it is up to DBA to establish and implement procedures for the use and maintenance of the archive.

Figure 8.7

Movement of order 12498 from live database to archive

Live Database

ORDERS

ORDNUMB	ORDDTE	CUSTNUMB
12489	90294	124
12491	90294	311
12494	90494	315
12495	90494	256
12498	90594	522
12500	90594	124
12504	90594	522

ORDLNE

ORDNUMB	PARTNUMB	NUMBORD	QUOTPRCE
12489	AX12	11	14.95
12491	BT04	1	402.99
12491	BZ66	1	311.95
12494	CB03	4	175.00
12495	CX11	2	57.95
12498	AZ52	2	22.95
12498	BA74	4	4.95
12500	BT04	1	402.99
12504	CZ81	2	108.99

Archive

ORDERS

ORDNUMB	ORDDTE	CUSTNUMB

ORDLNE

ORDNUMB	PARTNUMB	NUMBORD	QUOTPRCE

8.3   DATA DICTIONARY MANAGEMENT

In addition to administering the database, DBA also manages the **data dictionary**. The data dictionary is essentially the catalog mentioned in Chapter 7, but it often contains a wider range of information, including, at least, information on tables, columns, indexes, and programs.

DBA establishes naming conventions for tables, columns, indexes, and so on. It creates the data definitions for all tables as well as any data validation rules. It is also charged with the update of the contents of the data dictionary. The creation and distribution of appropriate reports from the data dictionary is another of its responsibilities.

8.4   TRAINING

DBA provides training in the use of the DBMS and in how to access the database. It also coordinates the training of users. In cases where training is provided by the vendor of software the organization has purchased, DBA handles the scheduling in order to make sure that the right users receive the training they require.

8.5   DBMS SUPPORT

DBMS Evaluation and Selection

DBA is responsible for the evaluation and selection of the DBMS. In order to oversee this responsibility, it sets up a checklist like the one shown in Figure 8.8. (This checklist applies specifically to a relational system, since this text deals with microcomputer DBMSs and most of these are, at least in part, relational. If we had not already selected a model for our focus, a category called "Choice of Data Model" would have to be added to the list. A corresponding checklist for mainframe systems would be slightly larger than the list shown in Figure 8.8. This example, however, is quite representative of a checklist for a relational DBMS.) DBA must evaluate each prospective purchase of a DBMS in terms of all the categories shown in the figure. An explanation of the various categories follows.

Figure 8.8

DBMS evaluation checklist

1. Data Definition
   a. Data types
      (1) Numeric
      (2) Character
      (3) Date
      (4) Logical (T/F)
      (5) Memo
      (6) Money
      (7) Other
   b. Support for nulls
   c. Support for primary keys
   d. Support for foreign keys
   e. Unique indexes
   f. Views

2. Data Restructuring
   a. Possible restructuring
      (1) Add new tables
      (2) Delete old tables
      (3) Add new columns
      (4) Change layout of existing
          columns
      (5) Delete columns
      (6) Add new indexes
      (7) Delete old indexes
   b. Ease of restructuring

3. Nonprocedural Languages
   a. Nonprocedural languages supported
      (1) SQL
      (2) QBE
      (3) Natural language
      (4) Own language. Award points on
          the basis of ease of use as well as
          the types of operations (e.g., joining,
          sorting, grouping, calculating various
          statistics) which are available in the
          language. SQL can be used as a standard
          against which such a language can be
          judged.
   b. Optimization done by one of the following:
      (1) User, in formulating the query
      (2) DBMS (through built-in optimizer)
      (3) No optimization possible. System will
          only do sequential searches.

4. Procedural Languages
   a. Procedural languages supported
      (1) Own language. Award points on the
          basis of the quality of this language
          both in terms of the types of statements
          and control structures available and
          the database manipulation statements
          included in the language.

Figure 8.8

(continued)

        (2) COBOL
        (3) FORTRAN
        (4) C
        (5) Pascal
        (6) BASIC
        (7) Other
    b. Can nonprocedural language be used in conjunction with the procedural language (e.g., could SQL be embedded in COBOL programs)?

5. Data Dictionary
    a. Types of entities
        (1) Tables
        (2) Columns
        (3) Indexes
        (4) Relationships
        (5) Programs
        (6) Other
    b. Integration of data dictionary with other components of the system

6. Shared Update
    a. Level of locking
        (1) Column
        (2) Row
        (3) Table
    b. Type of locking
        (1) Shared
        (2) Exclusive
        (3) Both
    c. Responsibility for handling deadlock
        (1) Programs
        (2) DBMS (automatic rollback of transaction causing deadlock)

7. Backup and Recovery Services
    a. Backup facilities
    b. Journaling facilities
    c. Recovery facilities
        (1) Recover from backup copy only
        (2) Recover using backup copy and journal
    d. Rollback of individual transactions

8. Security
    a. Passwords
        (1) Access to database only
        (2) Read or write access to any column or combination of columns
    b. Encryption
    c. Views
    d. Difficulty in bypassing security controls

9. Integrity
    a. Support for entity integrity
    b. Support for referential integrity
    c. Support for data-type integrity
    d. Support for other types of integrity constraints

10. Limitations
    a. Number of tables
    b. Number of columns
    c. Length of individual column
    d. Total length of all columns in a table
    e. Number of rows per table
    f. Number of files that can be open at the same time
    g. Types of hardware supported
    h. Types of LANs supported
    i. Other

11. Documentation
    a. Clearly written manuals
    b. Tutorial
        (1) Written
        (2) On-line
    c. On-line help available
        (1) General help
        (2) Context-sensitive help

12. Vendor Support
    a. Type of support available
    b. Quality of support available
    c. Cost of support
    d. Reputation of support

13. Performance
    a. Tests comparing the performance of various DBMSs in such areas as sorting, indexing, reading all rows, changing data values in all rows, and so on, are available from a variety of periodicals.
    b. If you have special requirements, you may want to design your own benchmark tests that could be performed on each DBMS under consideration.

14. Cost
    a. Cost of basic DBMS
    b. Cost of any additional components
    c. Cost of any additional hardware that is required
    d. Cost of network version (if required)
    e. Cost and types of support

15. Future Plans
    a. What does vendor plan for future of system?
    b. What is the history of the vendor in terms of keeping the system up-to-date?
    c. When changes are made in the system, what is involved in converting to the new version?
        (1) How easy is the conversion?
        (2) What will it cost?

16. Other Considerations (Fill in your own special requirements.)
    a. Special Purpose Reports
    b. ?
    c. ?
    d. ?

1. **Data definition.** What types of data are supported? Is support for nulls provided? What about primary and foreign keys? The DBMS will undoubtedly provide indexes, but is it possible to specify that an index is unique and then have the system enforce the uniqueness? Is support for views provided?

2. **Data restructuring.** What type of database restructuring is possible? How easy is it to do this restructuring? Will the system do most of the work or will the DBA have to create special programs for this purpose?

3. **Nonprocedural languages.** What type of nonprocedural language is supported? The possibilities are SQL, QBE, natural language, or a DBMS built-in language. If one of the standard languages is supported, how good a version is provided by the DBMS? If the DBMS furnishes its own language, how good is it? How does its functionality compare to that of SQL?

How does the DBMS achieve optimization of queries? Either the DBMS itself optimizes each query; the user must do so by the manner in which he or she states the query; or no optimization occurs. Most desirable, of course, is the first alternative.

4. **Procedural languages.** What types of procedural languages are supported? Are they common languages, such as COBOL, FORTRAN, BASIC, Pascal, or C, or does the DBMS come with its own language? In the latter case, how complete is the language? Does it contain all the required types of statements and control structures? What facilities are provided for accessing the database? Is it possible to make use of the nonprocedural language while using the procedural language?

5. **Data dictionary.** What kind of data dictionary support is available? Is it a simple catalog, or can it contain more, such as information about programs and the various data items these programs access? How well is the data dictionary integrated with other components of the system (for example, the nonprocedural language)?

6. **Shared update.** Is support provided for shared update? What is the unit that may be locked (column, row, or table)? Are exclusive locks the only ones permitted or are shared locks also allowed? (A shared lock permits other users to read the data; with an exclusive lock, no other user may access the data in any way.) How is deadlock handled? Will the DBMS take care of it, or is it the responsibility of programs to ensure that it is handled correctly?

7. **Backup and recovery services.** What type of backup and recovery facilities are provided? Can the DBMS maintain a journal of changes in the database and use the journal during the recovery process? If a transaction has aborted, is the DBMS capable of rolling it back (that is, undoing the updates of the transaction)?

8. **Security.** What type of security features does the system make available? Are passwords supported? Do passwords simply regulate whether a user may access the database, or is it possible to associate read or write access to a combination of columns with a password? Is encryption supported? Does the system have some type of view mechanism that can be used for security? How difficult is it to bypass the security controls?

9. **Integrity.** What type of integrity constraints are supported? Is there support for entity integrity (the fact that the primary key cannot be null)? What about referential integrity (the property where values in foreign keys must match values already in the database)? Does the DBMS support data-type integrity (the property where values that do not match the data type for the column into which they are being entered are not allowed to occur in the database)? Is there support for any other types of constraints?

10. **Limitations.** What limitations exist with respect to the number of tables, columns, and rows per table? How many files can be open at the same time? (Typically, each table and each index is in a separate file. Thus, a single table with three indexes, all in use at the same time, would account for *four* files. Problems may arise if the number of files that can be open is relatively small and many indexes are in use. On what types of hardware is the DBMS supported? What types of **local area networks (LANs)** support the DBMS?

(A local area network is a configuration of several computers all hooked together, thereby allowing users to share a variety of resources. One of these resources is the database. In a local area network, support for shared update is very important, since many users may be updating the database at the same time. The relevant question here, however, is not how well the DBMS supports shared update, but which of the LANs can be used in conjunction with this DBMS?)

11. **Documentation.** How good are the manuals? Are they easy to use? Is there a good index? Is a tutorial, in either printed or on-line form, available to assist users in getting started with the system? Is on-line help available? If so, is it general help or context-sensitive? (Context-sensitive help means that if a user is having trouble and asks for help, the DBMS will provide assistance for that particular problem at the time the user asks for it.)

12. **Vendor support.** What type of support is provided by the vendor, and how good is it? What is the cost? What is the vendor's reputation among current users?

13. **Performance.** How well does the system perform? This is a tough one to answer. One way to determine relative performance is to look into benchmark tests that have been performed on several DBMSs by various periodicals. Beyond this, if an organization has some specialized needs, it may have to set up its own benchmark tests.

14. **Cost.** What is the cost of the DBMS and any components the organization is planning to purchase? Is additional hardware required and, if so, what is the associated cost? If the organization requires a special version of the DBMS for a network, what is the additional cost? What is the cost of vendor support, and what types of support plans are available?

15. **Future plans.** What plans has the vendor made for the future of the system? This information is often difficult to obtain, but we can get an idea by looking at the performance of the vendor with respect to keeping the existing system up-to-date. How easy has it been for users to convert to new versions of the system?

16. **Other considerations.** This is a final, catch-all category that contains any special requirements not covered in the other categories.

Once each DBMS has been examined with respect to all the preceding categories, the results can be compared. Unfortunately, this process can be difficult, owing to the number of categories and their generally subjective nature. To make the process more objective, a numerical ranking can be assigned to each DBMS for its performance in each category (for example, a number between zero and ten, where zero is poor and ten is excellent). Further, the categories can be assigned weights. This allows an organization to signify which categories are more critical to it than others. Then each of the numbers being used in the numerical ranking can be multiplied by the appropriate weight. The results are added up, producing a weighted total. The weighted totals for each DBMS can then be compared, producing the final evaluation.

How does DBA arrive at the numbers to assign each DBMS in the various categories? Several methods are used. It can request feedback from other organizations that are currently using the DBMS in question. It can read journal reviews of the various DBMSs. Sometimes a trial version of the DBMS can be obtained, in which case members of the staff can give it a hands-on test. In practice, all three methods are sometimes combined. Whichever method is used, however, it is crucial that the checklist and weights be carefully thought out; otherwise, the findings may be inadvertently slanted in a particular direction.

Responsibility for the DBMS   Once the DBMS has been selected, DBA continues to be primarily responsible for it. DBA installs the DBMS in a way that is suitable for the organization. And, if the DBMS configuration needs to be changed, it is DBA that will make the changes.

When a new version of the DBMS is released, DBA will review it and determine whether the organization should convert to it. If the decision is made to convert to the new version (or perhaps to a new DBMS), DBA coordinates the conversion. Any fixes to problems in the DBMS which are sent by the vendor are also handled by DBA.

8.6 DATABASE DESIGN

DBA is responsible for carrying out the process of database design. It must ensure that a sound methodology for database design, such as the one discussed in Chapter 6, is established and is followed by all personnel who are involved in the process. It must also ensure that all pertinent information is obtained from the appropriate users.

DBA is responsible for the implementation of the final information-level design; in other words, it is responsible for the physical-level design process. If performance problems surface, it is up to DBA to make the changes that will improve the system's performance. This is called **tuning** the design.

DBA is also responsible for establishing standards for documentation of all the steps in the database design process. It also has to make sure that these standards are followed, that the documentation is kept up-to-date, and that the appropriate personnel have access to the documentation they need.

Requirements don't remain stable over time; they are constantly changing. DBA must review such changes and determine whether a change in the database design is warranted. If so, it must make such changes in the design and in the data in the database. It must also then make sure that all programs affected by the change are modified in any way necessary and that the corresponding documentation is also modified.

SUMMARY

1. Database administration (DBA) is the person or group that is assigned responsibility for supervising the database and the use of the DBMS.
2. DBA formulates and implements policies concerning the following:
   a. Those who can access the database; which portions of the database these persons may access, and in what manner.
   b. Security, that is, the prevention of unauthorized access to the database.
   c. Recovery of the database in the event that it is damaged.
   d. Management of an archive for data that is no longer needed in the database but must be retained for reference purposes.
3. DBA is in charge of maintaining the data dictionary.
4. DBA is in charge of training with respect to the use of the database and the DBMS. Training that is provided by an outside vendor is scheduled by DBA, which ensures that users receive the vendor training they need.
5. DBA is in charge of supporting the DBMS. This has two facets:
   a. The evaluation and selection of a new DBMS; DBA develops a checklist of desirable features for a DBMS and evaluates each prospective purchase of a DBMS against this list.
   b. DBA is responsible for installing and maintaining the DBMS after it has been selected and procured.
6. DBA is in charge of database design, both the information level and the physical level. It is also in charge of evaluating changes in requirements to determine whether a change in the database design is warranted. If so, DBA makes the change and reports it to affected users.

KEY TERMS

access privileges

archive

data archive

data dictionary

database administration
 (DBA)

disaster planning

integrity constraints

integrity support

local area network (LAN)

restructuring

tuning

EXERCISES

1. What is DBA? Why is it necessary?
2. What is DBA's role in regard to access privileges?
3. What is DBA's role in regard to security? What problems can arise in the use of passwords? How should these problems be handled?
4. Suppose a typical microcomputer DBMS is being used by company X. Suppose further that in the event the database is damaged in some way, it is essential that it be recovered *without the users having to redo any work*. What action should DBA take?
5. What are data archives? What purpose do they serve? What is the relationship between databases and data archives?
6. What is DBA's responsibility in regard to the data dictionary?
7. Who trains computer users within an organization? What is DBA's role in this training?
8. Describe the method that should be used to select a new DBMS.
9. For each of the following categories, what kinds of questions would DBA ask in order to evaluate a DBMS?
    a. Data definition
    b. Data restructuring
    c. Nonprocedural languages
    d. Procedural languages
    e. Data dictionary
    f. Shared update
    g. Backup and recovery services
    h. Security
    i. Integrity
    j. Limitations
    k. Documentation
    l. Vendor support
    m. Performance
    n. Cost
    o. Future plans
    p. Other considerations
10. How does DBA obtain the necessary information to award points for the various categories on the checklist?
11. What is DBA's role regarding the DBMS once it has been selected?
12. What is DBA's role in database design?

9

Application Generation

OBJECTIVES

1. Discuss the features offered by the typical application system.
2. Describe the components of an application generator.
3. Discuss the manner in which application generators are used to create an application system.

9.1 INTRODUCTION

When we use the term **application** or **application system**, we mean a software product, typically not just a single program, but a collection of programs, which is designed to satisfy certain *specific* needs of a user. A payroll system is a type of application system, as is a system that provides the functions of maintenance and reporting with regard to sales reps, customers, orders, and so on. The application system used by Lee provided maintenance and reporting with respect to directors, movies, movie stars, and so on. In this particular case, the user, Lee, wrote the application system himself. In other cases, it may be written by employees of the firm that will use it, or by programmers or software firms with which the user has contracted. In still other cases, it may be written by what is called a *software house*, which sells the same application system to a number of users.

The crucial idea is that an application system is written to handle a *specific* task. That task may be very complicated, but it is still something specific. A DBMS, on the other hand, is built to handle *general* types of tasks. It is up to the user of the DBMS to decide the manner in which the DBMS is used. In fact, DBMSs are used to write many, if not most, application systems.

Recently, a new type of software product, called an **application generator**, has emerged. This product, or tool, usually built around and closely integrated with a good DBMS, is designed to greatly improve the ease with which an application system can be developed. Other terms, such as fourth-generation language and fourth-generation environment, have also been applied to this highly valuable tool.

144

In section 9.2, we will investigate the features that are typically found in today's application systems. The components of application generators are covered in section 9.3. In section 9.4, we will discuss how these components are used to develop an application system. Finally, in section 9.5, we will discuss the other terms that are often used in place of the term application generator.

The focus of this chapter is microcomputer application systems and microcomputer application generators. Much of what will be said here applies equally well to mainframe environments, but they are more complex. For details about mainframes, see Chapter 13 of [14]. For a detailed discussion of how the advent of application generators has affected the whole system development process, see the same reference or [1].

9.2   APPLICATIONS

In what follows, we will attempt to discuss the characteristic **application system**. The features of application systems vary greatly; yet, they have enough in common to warrant a discussion of the "typical" system. You may encounter an application system that lacks one of the features we will mention. Such a system is by no means inferior because of this. It may well be that the feature in question is inappropriate for that particular system or that the system possesses some other feature that addresses the same need.

Menu

The typical application system is **menu-driven**; that is, the user indicates the action that he or she wishes to take by making the appropriate selection from a "menu" of options. A sample menu for a system for Premiere Products is shown in Figure 9.1. In this menu, users indicate their choice by entering the number associated with that option. In another menu style, individual options are highlighted. Through the use of arrow keys or a mouse, users highlight their selection and then press the return key or one of the buttons on the mouse. Figure 9.2 illustrates a menu with a highlighted box.

Figure 9.1

Menu for
*PREMIERE
PRODUCTS*

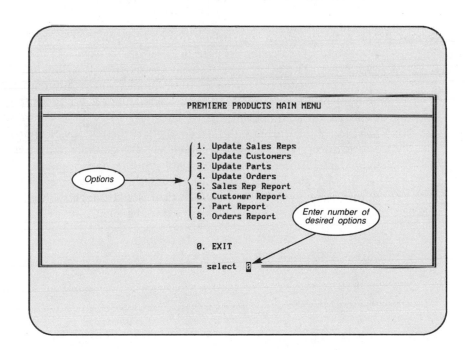

Figure 9.2

Menu for
*PREMIERE
PRODUCTS*

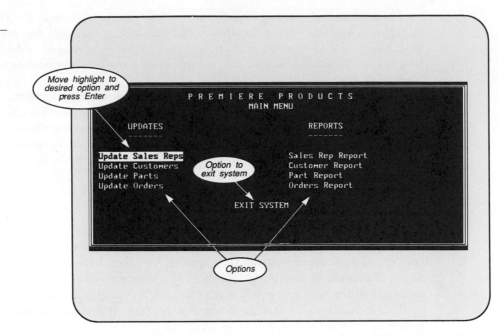

For some choices, the user's selection is now completed. The report will print or the update screen will be displayed, and the user can begin entering data. In other cases, another menu, called a submenu will appear and a second choice will be made in the same fashion as the first. For example, if the first choice was to update customer information, the second choice might be to (1) add, (2) change, or (3) delete a customer. Or, if the first choice was "PRINT REPORTS," the second choice might be to indicate which report was to be printed; whether the report was to go to the screen or the printer; whether to print the report in full detail or only the summary version; or whether all customers or only selected ones were to be printed.

Interactive and Automatic Updates

The systems usually support two types of updates. The most common one is the interactive update, so-called because the user interacts with the system by filling in forms on the screen. The other type is what we might call a "behind the scenes" or automatic update. A customer's balance, for example, might automatically be increased by the amount of an order when the order was shipped. Or fields that contain month-to-date totals would automatically be set back to zero at the end of a month. The user doesn't need to fill in a form on the screen to effect these kinds of updates.

Most updates are made interactively. For example, a **data entry screen** might be used to update customer data at Premiere Products (see Figure 9.3 on the next page). Users would indicate the type of action they wished to take by typing *A* for add, *C* for change, *D* for delete, or *S* for show. When they were through with all their updates, they would type *E* for end and would be returned to the menu.

Figure 9.3

Customer
Maintenance
screen

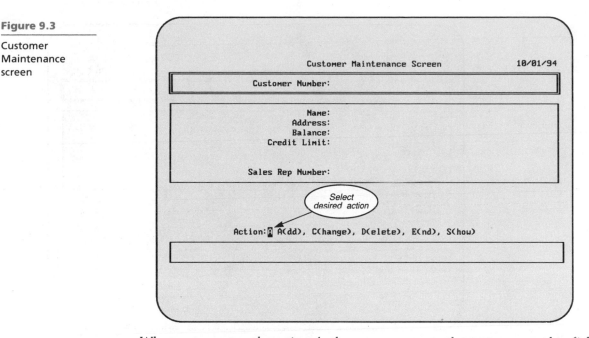

When a user enters the action *A*, the cursor moves to the customer number field and the system asks the user for a customer number. The system will first check to see whether a customer with that number is already filed in the database. If so, the user is given an error message (see Figure 9.4). If not, the user proceeds to fill in the rest of the data. If he or she makes any mistakes (for example, entering a sales rep number that is invalid), the system will give the appropriate error message (see Figure 9.5) and will force the user either to make the necessary correction or to abort the whole transaction.

Figure 9.4

Customer
Maintenance
screen

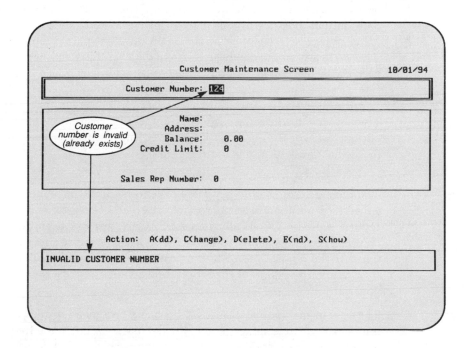

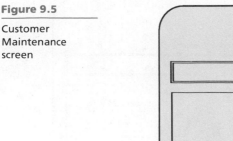

Figure 9.5

Customer
Maintenance
screen

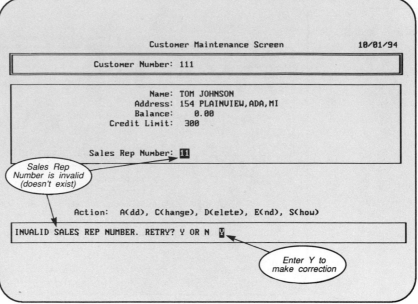

When all the data has been filled in, the system will give the user a chance to either accept the data in its present form or modify it before entering it into the database. Once the user has decided to accept the data as entered on the screen, it is added to the database.

The interaction is similar if the user chooses *C*. The only differences are that (1) the customer number that he or she enters must be the number of a customer who *is* already in the database and (2) current data for this customer is displayed on the screen before the user gets a chance to take further action. The user then steps through the form, entering data only in those fields which he or she wishes to change.

If a user selects *D*, he or she again must enter the customer number of a customer who is already in the database. Current data for this customer is then displayed. Then the user is asked whether he or she really wants to delete the customer. If the response is *Y* (yes), the customer is deleted. Otherwise, the customer is left in the database.

If a user wishes merely to view data on the screen, he or she selects the option *S*. Then he or she enters the customer number of a customer who is already in the database, current data for the customer is displayed, and the user is asked to press any key when he or she has finished looking at the data.

Reports

The typical system offers a variety of reports. Some of them are produced automatically as part of another process, for instance end-of-day or end-of-month routines. Other reports must be specifically requested. Once this has been done, users may be offered choices about how many copies to print, whether to use special paper to print the report, whether the report is to go to the printer or to the screen, whether to produce a detailed or a summary version of the report, and so on.

Queries

The system may support queries. By **query**, we mean a question asked by the user, the answer to which is somewhere in the database. Often, some frequently asked key queries are built into an application system; the user simply selects one from the menu and receives an answer. When we say that a system supports queries, however, we mean that users can ask these questions and get answers *whether or not the question has been pre-programmed*.

Backup and Recovery Services   The typical system provides a convenient way to do backup and recovery. In all probability, it will be the simplified approach discussed in Chapter 7; that is, a backup copy will be made periodically and the database will be recovered by copying the backup version over the live database. As noted earlier, this does entail redoing all work performed since the most recent backup.

Audit Trails   Audit trails are another service provided by the average application system. These are reports that detail whatever activity has taken place. When data is added to the database, the audit trail shows the new data. When it is deleted, the audit trail shows how the data looked *before* the deletion. When data is changed, the audit trail shows how the data looked *before* and *after* the change. The audit trail enables users to ensure the accuracy of updates, and, if a mistake is detected, the data reported in the audit trail aids the user in correcting it.

Utility Services   The typical system also provides utility services. One such service maintains statistics on types of activities users are engaged in and periodically reports on these statistics. Other services periodically check the data in the database for errors (for example, a customer with a sales rep who is not in the database) and print a report describing any errors that have been found.

Administrative Services   The typical application system provides administrative services. These are the services that would be utilized by the person or group (DBA) that is in charge of the system, and they include assigning passwords to new users, changing passwords, specifying whether other security features, such as encryption, are to be used; and varying any of the system's parameters to make the system run more smoothly or to more closely model the workings of the organization. Backup and recovery services, along with some of the utility services, could fall into this category as well.

The features listed in this section are summarized in Figure 9.6.

Figure 9.6

Features of a typical application system

Features of a Typical Application System

1. Menu
2. Interactive and Automatic Updates
3. Reports
4. Queries
5. Backup and Recovery Services
6. Audit Trails
7. Utility Services
8. Administrative Services

9.3 APPLICATION GENERATORS: COMPONENTS

Anyone who has ever been involved in the development of an application system knows that it is a difficult and complicated job, much of which is tedious and repetitive. For example, writing a program in BASIC that employs a simple form on the screen for the interactive updating of customer data is not an easy process; it involves the careful manipulation of many nitty-gritty details. Even putting the form on the screen is tedious. For each piece of the form, the cursor must be moved to the appropriate position on the screen, and then the appropriate characters must be written. For each of the data fields on the screen (the positions where the users must enter data), the cursor must be moved to the correct position and the data must be read and validated. Since the data entered may not be valid, the program must supply a mechanism for handling errors.

Many program statements are required to accomplish these tasks and an application system consists of many programs. Every data entry program in the system must contend with all these details. Creating a program like the one we just discussed is not necessarily difficult, but it's usually time consuming. And until the program has been completed, the user can't really see what the form looks like, that is, how it will appear on the screen. When the user does finally see the form, he or she may or may not like it! If the form is not acceptable to the user, major changes may have to be made to the program.

Clearly, it would be very helpful to have a much simpler way to create data entry programs and all the other types of programs that go into an application system. This is what an application generator is all about. It furnishes an alternate and more productive way to design and develop the features of an application system.

An **application generator** is a many-faceted software product. As you read the following about the various facilities an application generator should offer (see Figure 9.7), keep in mind the goal: the efficient development of application systems.

Figure 9.7

Features of an application generator

Features of an Application Generator

1. Programmer's Workbench
2. Data Dictionary
3. DBMS, Preferably Relational
4. Screen Painter (or Screen Generator)
5. Query Facility
6. Report Writer
7. Nonprocedural Language
8. Procedural Language
9. Menu Generator
10. Help Facility
11. Program Generator

Programmer's Workbench

If you're handier than I am, you probably have a home workbench. It contains the tools you need in order to work efficiently. A **programmer's workbench** is similar, in that it contains tools that simplify various tasks, in this case, those tasks which are necessary to develop an application system. It contains an efficient, easy-to-use editor that provides for the rapid entry and modification of programs. In addition, it contains several utilities for compiling, debugging, and producing reports that are crucial during the application-development process.

Data Dictionary

The **data dictionary** contains descriptions of all the data items in the system. At the very least, it contains the description of all the tables in the system as well as the names and physical characteristics of all the columns that make up these tables. Ideally, it should also contain any indexes that are associated with the tables as well as a description of any relationships between the tables. Information in the dictionary is used by the other components of the application generator.

DBMS, Preferably Relational

To realize all the benefits conferred by the database approach to processing, the application generator should be geared to work with a database rather than with collections of unrelated files. To manage such a database requires a good DBMS, and to achieve the maximum flexibility, the DBMS should be relational. What we are saying here is that a good relational DBMS should be at the heart of the application generator. In essence, the application generator is built around this DBMS.

Screen Painter   A **screen painter**, also called a **screen generator**, is a tool that is used to facilitate the development of screen-oriented data-entry programs. Various screen painters approach the task differently, but they all have one thing in common: to use them, we merely sit down at the keyboard and indicate in some simple way exactly how a form is to appear. As we are doing this, the system displays the form as far as we have constructed it, so we continually get visual feedback on what we are doing. It feels as though we are "painting" the form on the screen and this is how the tool got its name.

The methods of screen painters vary; the one we'll discuss is typical. First, we are presented with a blank screen. To describe a literal part of the screen (that is, a specific string of characters like "Customer Number:"), we use the arrow keys to move the cursor to the correct location and then type the appropriate characters. Figure 9.8 illustrates what the screen might look like at this point. Now we are ready to describe the rest of the **background** (that is, the portion of the form that does not involve data). Figure 9.9 illustrates a completed background.

Figure 9.8

Screen painter

Figure 9.9

Screen painter

Once the background looks the way we want it to, we can describe the foreground (that is, the data portion of the form). To create the first field in the foreground, we move the cursor to the spot where this field begins. At this point, we are asked to describe the characteristics of the data that is to be entered. Is it a numeric field, a character field, a date, or some other type? How long is it? If it is numeric, how many decimal places does it have? Does it have any special layout (for example, a social security number that must be in the form 999-99-9999)?

There are two commonly used ways of doing this. If the field is a column in the database (like the *CUSTNUMB* column in the *CUSTOMER* table) which has already been described to the data dictionary, all the sought-after information is already in the dictionary, and simply indicating the column's name may be sufficient. If this is not the case, then we may have to supply direct answers to all the questions. Either way, once this has been done, we are finished with the first field in the foreground and can move on to the next.

We may also wish to engage special effects, such as highlighting (making the data in the field brighter), reverse video (dark letters on a light background instead of light letters on a dark background), blinking, or the use of special colors. When this has been done for the *CUSTNUMB* field, the form may look something like the one shown in Figure 9.10. In order to complete the process, we specify each of the data fields that make up the foreground portion of the form. We also might add other features, such as boxes, to enhance the appearance of the form. When completed, the form might look something like the one in Figure 9.11.

Figure 9.10

Screen painter

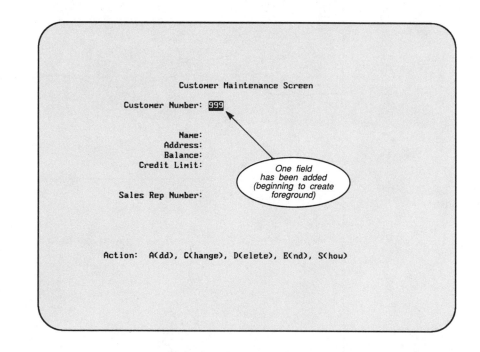

Figure 9.11

Screen painter

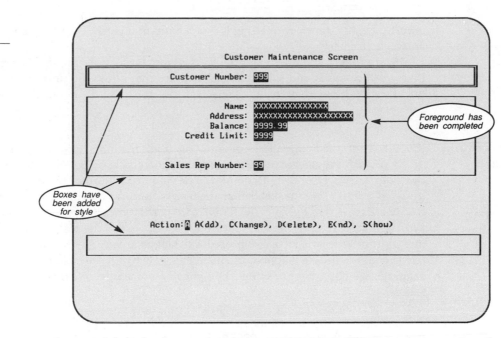

Any one of the background or foreground fields can be moved at any time if we feel it is not in the right position. Some systems accomplish this by having us place the cursor at the first position of the field where it is currently located, hit a special key, move the cursor to the first position of the spot on the screen where we would like to locate the field, and hit another key. The system then completes the work of moving the field.

Query Facility

A **query facility** is a tool that allows queries to the database to be answered in a simple fashion. We may use this facility in developing the application system. For instance, it can be used to build into the application system those queries which we can anticipate in advance. Users are also apt to want to query the database in ways that have not been anticipated (these are called ad hoc queries). Without the use of the query facility, these users would either have to write programs themselves or, more likely, turn in a request to the programming staff to get a program written for them.

Application generators use various types of query facilities. In some, the user types a brief command in a language like **SQL**, which is becoming quite common in such an environment. In others, the user fills in forms by means of a method that is similar to **QBE**. In still others, the user answers a series of questions to indicate his or her queries. **Natural languages** are yet another option. The main thing held in common by all of these approaches is that they provide a simple way for the nontechnical user to specify his or her query.

Report Writer

A **report writer** is similar to a query facility in that it simplifies; specifically, it is intended to facilitate the process of developing reports. It enables the user to describe quickly and easily what a report is to look like; conditions that determine what is to be included in the report; page headings; column headings; totals and subtotals; and so on. We can then use this description in our application. Whenever the user of the application chooses this report, a report will be printed whose layout matches the one that has been described. Further, only the data that satisfies established criteria is included.

Given their similarity, what is the difference between a report writer and a query facility? The line between them is fuzzy at best. When are we submitting a query to which we seek a reply, and when are we requesting a report of all data that meets certain conditions? The answer is not totally clear, and there is a large area of overlap. If we are going to make a distinction, we could base it on three things. First, the answer to a query is usually quite brief; a report is customarily longer. Second, answers to queries are displayed on a screen more often than they are printed, whereas the opposite is true for reports. Finally, fancy formatting is much more of a consideration in a report than in the answer to a query.

Many query facilities can double as report writers. To do so, they include features that produce totals and subtotals and facilities that allow the flexible formatting of output. Provided that they include these features, they can be included in the report writer category.

Nonprocedural Language

If you have studied computer programming at all, the language you studied was probably in the procedural category. With procedural languages, as noted earlier, we must tell the computer precisely *how* to accomplish a desired task. In other words, the computer must be given a *procedure* (also called an *algorithm*) to follow in order to attain the objective. BASIC, Pascal, COBOL, FORTRAN, and many other languages are in this category. With a nonprocedural language, we only specify *what* the task is. In this text, the primary prime example of a nonprocedural language is SQL. If we want to use SQL to retrieve all the orders placed at Premiere Products by all the customers of sales rep 12, for example, we don't have to write a complicated procedure. We merely indicate the task as follows:

```
SELECT ORDNUMB, ORDDATE
    FROM CUSTOMER, ORDERS
    WHERE CUSTOMER.ORDNUMB = ORDERS.ORDNUMB
    AND SLSRNUMB = 12
```

Then it is up to the computer to figure out *how* to get the job done. Clearly, if the computer takes over this responsibility, we can be much more productive.

Procedural Language

Why do we need a nonprocedural language to be available as part of the application generator? The answer is obvious: it makes us much more productive in the development of an application. Given this, why do we need a procedural language to be available as well? The reason is that nonprocedural languages can't do everything. Some of them are very powerful and are capable of handling a wide variety of tasks, but, no matter how powerful they are, we are still likely to encounter tasks that are beyond their capabilities. In such cases, we need to be able to resort to some good procedural language.

This isn't the whole story, however. We need the two languages, the nonprocedural and the procedural, to be closely integrated. In some environments that furnish both types of languages, once the task is beyond the capabilities of the nonprocedural language, that language becomes totally useless. Our only choice is to turn to the procedural language and give up all the benefits offered by its nonprocedural counterpart.

Ideally, commands from the nonprocedural language can be embedded in a program written in the procedural language. Then both languages can serve a role: the nonprocedural language handles everything it can, and the procedural language handles only those tasks which are left over. Many application generators furnish this good, close relationship between the two languages.

Menu Generator

Some application generators include a facility to assist us in putting all these pieces together in a convenient menu-driven environment. Since the application system we are developing ought to be menu-driven (because this is easiest for the user), such a facility is very important. Without it, we have to write programs to display menus, get choices from users, and take appropriate actions.

Help Facility

A **help facility** is simply a vehicle whereby users can obtain information about the system they are using without having to resort to a printed manual. Typically, a user presses a special key (Function key 1 is commonly used for this), and when he or she does, information about the part of the system he or she is using appears on the screen. After the user has viewed this information, he or she hits another key and resumes working with the system at the point where he or she left off.

When we talk about an application generator, we refer to two different help facilities. First, the application generator itself should have a help facility that can immediately provide us with information concerning the use of the application generator any time we seek it. Second, we would like to build a help facility into our application systems. Some application generators include facilities to assist us in doing this. In the absence of such facilities, we have to come up with the code to furnish this help to the users of our application system. This is not especially difficult to do, but it does take time; so it is far preferable for the application generator itself to possess this facility.

Program Generator

Basically, application generators work in one of two ways. Let's use the report writer facility to illustrate the difference. In the first approach, we describe a report; when the report is requested and printed, it will match the layout we have described. Sometime later, the report layout can be changed, and the next time the report is printed, it will match the new specifications. The drawback here is that we can't access the report description except through the report writer facility; in other words, this tool is essential for making changes in the report. This is reminiscent of the difficulty of working only with a nonprocedural language. If the report is beyond the ability of the report writer to produce, the tool is of no value to us; we have to write a complete program to do the job, just as if we had no report writer to assist us at all.

In the second approach, we also describe the layout of the report to the report writer, but then the report writer *creates a program which would be run to produce the report*. At this point, if we wanted to, we could make modifications in this program. We could still use such a report writer, even if the report to be produced is beyond its scope. We would specify a report that had as many of the characteristics we wanted as the report writer could handle. At this point, we would let the report writer create the program. Then we would make whatever changes were necessary in this program so that it would do the things that were beyond the scope of the report writer. Note that as with the first approach, a special program still has to be written. The difference here is that the report writer will create a large share of it and the portion left for us to write, we hope, is quite small.

This approach is often called a **program generator**. It actually creates, or *generates*, programs for us. As you can see, the benefits are considerable. Note that although we used the report writer feature to illustrate the concept, its application is not limited to that. The screen painter and the menu generator are examples of other facilities to which it is quite well suited.

9.4 APPLICATION GENERATORS: UTILIZATION

How do we utilize all these components to put together an application system? First, we need to determine the various reports, queries, and updates that make up the application system, and we need to design the underlying database. Then we can begin to use the application generator. (Actually, some of the work in the first step can proceed in parallel with some of the work with the application generator, but we will assume that the work in the first step has already been completed.)

We begin using the application generator by describing the contents of the database to the data dictionary. We indicate the columns that comprise each table, and the characteristics of these columns. For the *CUSTOMER* table in the *PREMIERE PRODUCTS* application system, for example, we might enter the data shown in Figure 9.12 on the next page. The other tables that make up the *PREMIERE PRODUCTS* database would be described in a similar fashion. If the system also allows us to describe relationships between tables, we will do so.

Figure 9.12

Layout of
CUSTOMER table

Structure for CUSTOMER table

Field	Field Name	Type	Width	Dec
1	CUSTNUMB	Numeric	3	
2	CUSTNAME	Character	15	
3	ADDRESS	Character	20	
4	BALANCE	Numeric	7	2
5	CREDLIM	Numeric	4	
6	SLSRNUMB	Numeric	2	

** Total **   52

Dec stands for decimal places, and the total includes a
decimal point in the *BALANCE* field.

After these descriptions have been entered, we use the screen painter to create the forms for all the on-line updates. After the forms have been created, we can begin to enter sample data. (Many systems automatically provide a simple form that can be used to actually enter some of the data even before these update forms have been created. An example of such a form is shown in Figure 9.13.)

Figure 9.13

System-created
update form for
CUSTOMER table

```
CUSTNUMB    999
CUSTNAME    XXXXXXXXXXXXXXX
ADDRESS     XXXXXXXXXXXXXXXXXXXX
BALANCE     9999.99
CREDLIM     9999
SLSRNUMB    99
```

We then move on to any reports and/or queries that are required by the system, and we use the report writer and query facility to create them. Since we have already entered some sample data, we can test these reports and queries as soon as they are created.

Whatever is beyond the scope of the screen painter, report writer, and/or query facility must now be added to the system. This will probably mean writing commands in either the nonprocedural or procedural language, either by adding commands to programs generated by the system or by writing entire programs from scratch. We always attempt to use the nonprocedural language first. Only when we encounter something that is beyond the capabilities of the nonprocedural language do we use the procedural language. Even then, we retain as much as we can of what has been accomplished with the nonprocedural language.

Programs also have to be written for functions that are not addressed by the application generator. This typically includes utility programs, automatic updates, and backup and recovery programs.

Finally, everything must be tied together in a menu-driven environment. We can do this either after we've prepared all the individual pieces or as they are being developed. Many application generators have facilities for developing menu-driven systems. If the one we are using does not, a program must be written to accomplish the same task.

Including a help facility is an option. As in other cases, if the application generator contains features that can assist us in developing this help facility, we will use them; if not, we must write the code to implement such a facility.

The steps involved in developing an application are summarized in Figure 9.14 on the next page.

Figure 9.14	**Steps in Developing an Application**
Steps in developing an application system with an application generator	1. Describe contents of the database to the data dictionary.
	2. Use screen painter to create forms for on-line updates.
	3. Enter sample data.
	4. Create reports and/or queries. Test with sample data.
	5. Add missing features, using nonprocedural language wherever possible. Use procedural language for anything else.
	6. Create menus for all of the choices. Use menu-generation facility of the application generator if it has one; otherwise, write a menu program.
	7. Consider including help facility. Again, use the tools provided by the application generator to do this if one is present. If not, modify programs you have written to include this facility.

9.5 APPLICATION GENERATORS: OTHER TERMINOLOGY

Unfortunately, there is no general agreement on terminology here. Some people use the term application generators, as we have done in this chapter. Others use the term **fourth-generation language (4GL)**. (Machine languages comprised the first generation of computer languages. Assembly language, a major advance in productivity for programmers, marked the second generation. The so-called *high-level languages* like FORTRAN, COBOL, BASIC, and Pascal, are part of the third generation; they provided another major increase in productivity.)

Since application generators made programmers still *more* productive, the term fourth generation language is often applied to them. However, the term is used inconsistently. Some people use it to refer only to the nonprocedural part of an application generator, whereas others use it to refer to all its components; that is, to the whole "environment." This second usage has led to another term, the **fourth-generation environment**. This refers to an environment that is inclusive of all the various features we discussed with regard to application generators.

Thus, the terms application generator and fourth-generation environment are synonymous. Whether the terms application generator and fourth-generation language are synonymous depends on whom you ask.

One more note of interest on this subject. There is another definition for fourth-generation language which is even harder to get a handle on. Some people say it is any language or environment that makes a programmer ten or more times more productive than a third-generation language would allow him or her to be. While this definition accurately reflects the goal of a fourth-generation language, it is certainly not very easy to apply!

SUMMARY

1. An application system is a system that has been developed to handle some specific task. A microcomputer application system typically provides the following:

   a. A menu, from which users select the action to be taken.

   b. Interactive updates through forms on the screen that are filled in by the user (some of the updates may also be made automatically).

   c. A variety of reports.

   d. A simple way for users to query the database.

   e. Facilities for backup and recovery.

   f. Audit trails, that is, records of all the updates that have taken place in the database.

   g. A range of utility-type services.

   h. Various administrative-type services.

2. Application generators are tools that are used for the rapid development of an application system. The components of such tools are as follows:
   a. A programmer's workbench, which includes the various tools a programmer needs in order to use the other components of the system effectively.
   b. A data dictionary, which contains descriptions of the tables and columns that make up the database and which is integrated with the other components of the system.
   c. A DBMS, which manipulates the data.
   d. A screen painter, which facilitates the development of screen-oriented on-line update programs.
   e. A query facility, which provides a simple mechanism for querying the database.
   f. A report writer, which facilitates development of the reports required in the application system.
   g. A nonprocedural language, that is, one in which we specify *what* the task is rather than *how* it is to be done.
   h. A procedural language, which handles the tasks that are beyond the scope of the nonprocedural language; the two languages should be able to work together.
   i. A menu generator, which allows us to link the various operations in the application system with a menu of available choices.
   j. Two types of help facilities, one that gives us help on the application generator as we are using it, and one that provides us with a simple means of including a help facility of our own in the application system we are developing.
   k. A program generator, which creates programs we can then modify, if necessary, to accomplish a given task.
3. Some people use the term fourth-generation language as a synonym for application generator; others use the term to apply only to the nonprocedural language component. Another term for application generators is fourth-generation environment.

KEY TERMS

application	nonprocedural language
application generator	procedural language
application system	program generator
data dictionary	programmer's workbench
fourth-generation environment	query
fourth-generation language (4GL)	report writer
help facility	screen generator
menu generator	screen painter

EXERCISES

1. What is meant by an application system? Who creates application systems? Is a DBMS an application system?
2. What is an application generator? How does it differ from an application system?
3. List and briefly describe the components of a modern application system.
4. What do we mean when we say that an application system is menu-driven?
5. How are interactive updates usually achieved in a modern application system? Are any updates noninteractive?
6. Describe the type of backup and recovery procedures that are provided by a typical microcomputer application system. What problems, if any, are associated with these procedures?
7. What is an audit trail? What is its purpose?

8. What types of administrative services might be furnished as part of a microcomputer application system?
9. Describe a programmer's workbench.
10. What is a data dictionary? What role does it play in an application generator?
11. What is a screen painter? How is it used? What is another name for it?
12. What is a query facility? What is a report writer? Can you describe a way in which they differ?
13. What is a nonprocedural language? Give an example. What purpose does it serve?
14. What is a procedural language? Give an example. Explain its usefulness.
15. Does an application generator need both a nonprocedural language and a procedural language? If an application generator contains both types of languages, what relationship should exist between them?
16. What is a menu generator? What purpose does it serve? What action should be taken if an application generator does not contain one?
17. What is a help facility? What purpose does it serve? What should be done if an application generator does not contain one?
18. What is a program generator? Is it essential for an application generator to contain one? What advantage does a program generator offer?
19. List the steps involved in developing an application system by means of an application generator.
20. Describe the relationship between the following terms: application generator, fourth-generation language, fourth-generation environment.

Answers to Odd-Numbered Exercises

CHAPTER 1 — INTRODUCTION TO DATABASE MANAGEMENT

1. A software package is a collection of programs that is designed to handle some specific task, such as payroll. It is also called a software system, an application system, or an application package.

3. A file is a structure that is used to store data about a single entity; it can be viewed as a table. A record is a row in the table. A field is a column.

5. The number of entities and complex relationships, along with the fact that the entire billing operation of the practice would depend on the system, led Pat to conclude that the problem should be put in the hands of computer professionals.

7. A relationship is an association between entities.

9. A database is a structure that can house information about several types of entities, the attributes of these entities, and the relationships among them.

11. The purchase price of microcomputer DBMSs is relatively low, ranging from $100 to $800. In contrast, the cost of a good mainframe DBMS can run as high as $400,000.

13. Sharing of data means that many users will have access to the same data.

15. Redundancy is the duplication of data. It wastes space, makes updating more difficult, and may lead to inconsistencies in the data.

17. Integrity means that the data in the database follows certain rules (called integrity constraints) that users have established.

19. Data independence is the property by which the structure of a database can be altered without changes having to be made in the programs that access the database. With data independence, it is easy to change the structure of the database when the need arises.

21. The more complex a product is in general (and a DBMS, in particular, is complex), the more difficult it is to understand and correctly apply its features. As a result of this complexity, serious problems may result from mistakes made by the user of the DBMS.

23. Recovery can be more difficult in a database environment, partly because of the greater complexity of the structure. It is also likely that in the database environment several users will be making updates at the same time, which means that recovering the database involves not only restoring it to the last state in which it was known to be correct, but also performing the complex task of redoing all the updates made since that time.

CHAPTER 2 — DATA MODELS

1. A data model is a category of the database management system. It has two components: structure and operations. *Structure* refers to the format in which the DBMS stores data, and *operations* refers to the facilities that are given to users for the purpose of manipulating the data.

3. The three main data models are the relational model, the network model, and the hierarchical model.

5. Order lines in a separate table create a simpler structure, since only one entry is made in any box in the table. Processing that involves finding all the order lines for a given part is much simpler, since a maximum number of order lines does not have to be determined in advance, as it would be if order lines had to be kept in the *Orders* table. Further, no space is wasted by orders that have very few order lines.

7. A relational database is a collection of tables (relations).

9. A tuple is the formal name in the relational model for row; thus, another name for tuple is *row*. It also corresponds to the term *record*.

11. In the shorthand representation, each table is listed, and after each table, all the columns of the table are listed in parentheses. Primary keys are underlined.

```
BRANCH (BRNUMB, BRNAME, BRLOC, NUMEMP)
PUBLSHR (PUBCODE, PUBNAME, PUBCITY)
AUTHOR (AUTHNUMB, AUTHNAME)
BOOK (BKCODE, BKTITLE, PUBCODE, BKTYPE, BKPRICE, PB)
WROTE (BKCODE, AUTHNUMB, SEQNUMB)
INVENT (BKCODE, BRNUMB, OH)
```

13. The primary key is the column or collection of columns that uniquely identifies a given row. The primary key of the *Publshr* table is *Pubcode*. The primary key of the *Author* table is *Authnumb*. The primary key of the *Book* table is *Bkcode*. The primary key of the *Wrote* table is the concatenation (combination) of *Bkcode* and *AUTHNUMB*.

15. The relational model systems are easier to use than systems that follow the other models. Changing the structure of a database is also easier in relational model systems. The disadvantages of relational systems are that they are less efficient than systems that follow the other models, and they provide less support for integrity.

17. The CODASYL model is properly a subset of the network model. To many people, however, the two have become synonymous.

19. Since the network model has a special structure to effect relationships, there is no need for the matching columns that are required by the relational model.

21. Network model systems are more efficient than relational systems, and they provide better support for some types of integrity constraints. The disadvantages are that they are harder to use and they furnish less data independence than relational systems.

23. For any child that has more than one parent, each parent must be in a different physical database. For any parent that is not in the same physical database as the child, the relationship is a logical child relationship.

CHAPTER 3 — THE RELATIONAL MODEL: DATA DEFINITION AND MANIPULATION

(For Exercises 1 through 17, only the SQL formulations are shown.)

1.

```
SELECT PARTNUMB, PARTDESC
    FROM PART
```

3.

```
SELECT CUSTNAME
      FROM CUSTOMER
      WHERE CREDLIM >= 800
```

5.

```
SELECT PARTNUMB, PARTDESC,
      UNONHAND * UNITPRCE
      FROM PART
      WHERE ITEMCLSS = 'AP'
```

7.

```
SELECT *
      FROM PART
      ORDERS BY UNITPRCE
```

9.

```
SELECT SUM(BALANCE)
      FROM CUSTOMER
      WHERE SLSRNUMB = 12
```

11.

```
SELECT ORDNUMB, DATE, CUSTOMER.CUSTNUMB,
      CUSTNAME
      FROM CUSTOMER, ORDERS
      WHERE CUSTOMER.CUSTNUMB = ORDERS.CUSTNUMB
      AND DATE = 90594
```

13.

```
SELECT ORDNUMB, DATE, CUSTOMER.CUSTNUMB,
      CUSTNAME, SLSREP.SLSRNUMB, SLSRNAME
      FROM SLSREP, CUSTOMER, ORDERS
      WHERE CUSTOMER.CUSTNUMB = ORDERS.CUSTNUMB
      AND SLSREP.SLSRNUMB = CUSTOMER.SLSRNUMB
```

15.

```
INSERT INTO ORDERS
      VALUES
      (12600, 90694, 311)
```

17.

```
CREATE TABLE SPGOOD
      (PARTNUMB        CHAR(8),
       PARTDESC        CHAR(25),
       UNITPRCE        DECIMAL (6,2))
INSERT INTO SPGOOD
      SELECT PARTNUMB, PARTDESC, UNITPRCE
            FROM PART
            WHERE ITEMCLSS = 'SG'
```

19.

CUSTOMER	CUSTNUMB	CUSTNAME	ADDRESS	etc.	
P.					

21.

CUSTOMER	CUSTNUMB	CUSTNAME	CREDLIM	SLSRNUMB	
	P.	P.	500	3	

23.

CUSTOMER	CUSTNUMB	CUSTNAME	SLSRNUMB	etc.	
	P.	P.	~3		

25. Relational algebra:

```
SELECT PART WHERE PARTNUMB = 'BT04' GIVING ANSWER
```

SQL:

```
SELECT *
    FROM PART
    WHERE PARTNUMB = 'BT04'
```

27. Relational algebra:

```
JOIN CUSTOMER ORDERS
    WHERE CUSTOMER.CUSTNUMB = ORDER.CUSTNUMB
    GIVING TEMP
PROJECT TEMP OVER (ORDNUMB, ORDDTE, CUSTNUMB,
    CUSTNAME) GIVING ANSWER
```

SQL:

```
SELECT ORDNUMB, DATE, CUSTOMER.CUSTNUMB, CUSTNAME
    FROM CUSTOMER, ORDERS
    WHERE CUSTOMER.CUSTNUMB = ORDERS.CUSTNUMB
```

CHAPTER 4 — RELATIONAL MODEL II: ADVANCED TOPICS

1. A view is an individual user's picture of the database. It is defined through a defining query. The data in the view never actually exists in the form described in the view. Rather, when a user accesses the view, his or her query is merged with the defining query of the view to form a query that pertains to the whole database.

3. a.
```
CREATE VIEW CUSTORD AS
      SELECT CUSTOMER.CUSTNUMB, CUSTNAME,
            BALANCE, ORDNUMB, DATE
            FROM CUSTOMER, ORDERS
            WHERE CUSTOMER.CUSTNUMB =
                  ORDERS.CUSTNUMB
```

b.
```
SELECT CUSTNUMB, CUSTNAME, ORDNUMB, DATE
      FROM CUSTORD
      WHERE BALANCE > 100
```

c.
```
SELECT CUSTOMER.CUSTNUMB, CUSTNAME, ORDNUMB,
      DATE
      FROM CUSTOMER, ORDERS
            WHERE CUSTOMER.CUSTNUMB =
                  ORDERS.CUSTNUMB
            AND BALANCE > 100
```

5. On relational mainframe DBMSs, the optimizer (a part of the DBMS) makes the decision to use a particular index. On microcomputer DBMSs, the user or programmer makes this decision.

7. If the DBMS updates the catalog automatically, users need not worry about having to update the catalog whenever they make a change in the database structure. If they did have to make such a change, it might be made incorrectly, in which case the data in the catalog wouldn't match the structure of the database.

9. The structure of a table can be changed in SQL through the ALTER command. Columns can be added (ALTER TABLE table-name ADD column-name); columns can be deleted (ALTER TABLE table-name DELETE column-name); and columns can be changed (ALTER TABLE table-name CHANGE COLUMN column-name TO new description). Tables can be deleted (DROP TABLE table-name).

11. A tabular system is one in which users perceive databases as collections of tables. A minimally relational system is one in which users perceive databases as collections of tables and which supports the SELECT, PROJECT, and JOIN commands of the relational algebra. A relationally complete system is one in which users perceive databases as collections of tables and which supports the complete relational algebra. A fully relational system is one in which users perceive databases as collections of tables and which supports the complete relational algebra as well as entity and referential integrity.

CHAPTER 5 — DATABASE DESIGN I: NORMALIZATION

1. Column B is functionally dependent on column A if a value for A determines a unique value for B at any time.

3. The primary key of a table is the column or collection of columns that determines all other columns in the table and for which there is no subcollection that also determines all other columns.

5. A table is in first normal form if it does not contain a repeating group.

7. A table is in third normal form if it is in second normal form and if the only determinants it contains are candidate keys. If a table is not in 3NF, redundant data will cause wasted space and update problems. Inconsistent data may also be a problem.

9. Many answers are possible. See the guidelines in Chapter 5.

11.

```
INVNUMB --> CUSTNUMB, CUSTNAME,
      CUSTADDR, INVDATE
CUSTNUMB --> CUSTNAME, CUSTADDR
PARTNUMB --> PARTDESC, PRICE
INVNUMB, PARTNUMB --> NUMBSHIP


INVOICE (INVNUMB, CUSTNUMB, INVDATE)
CUSTOMER (CUSTNUMB, CUSTNAME, CUSTADDR)
PART (PARTNUMB, PARTDESC, PRICE)
INVLNE (INVNUMB, PARTNUMB, NUMBSHIP)
```

CHAPTER 6 — DATABASE DESIGN II: DESIGN METHODOLOGY

1. A user view is the view of data that is necessary to support the operations of a particular user. By considering individual user views rather than the complete design problem, we greatly simplify the database design process.

3. If the design problem were extremely simple, the overall design might not have to be broken down into a consideration of individual user views.

5. The primary key is the column or columns that uniquely identify a given row and that furnish the main mechanism for directly accessing a row in the table. An alternate key is a column or combination of columns that could have functioned as the primary key but was not chosen to do so. A secondary key is a column or combination of columns that is not any other type of key but is of interest for purposes of retrieval. A foreign key is a column or combination of columns in one table whose values that are required to match the primary key in another table. Foreign keys furnish the mechanism through which relationships are made explicit.

7. a. Include the project number as a foreign key in the employee table.
   b. Include the employee number as a foreign key in the project table.
   c. Create a new table whose primary key is the concatenation of employee number and project number.

9. Instead of the advisor number being included as a foreign key in the student table, there would be an additional table whose primary key was the concatenation of student number and advisor number.

11.

```
BRANCH (BRNUMB, BRNAME, BRLOC, NUMEMP)
      Primary key BRNUMB
      Secondary key BRNAME
      1.   BRNUMB must be unique.

PUBLSHR (PUBCODE, PUBNAME, PUBCITY)
      Primary key PUBCODE
      Secondary key PUBNAME
      1.   PUBCODE must be unique.
```

```
AUTHOR(AUTHNUMB, AUTHNAME)
     Primary key AUTHNUMB
     1.    AUTHNUMB must be unique.

BOOK (BKCODE, BKTITLE, PUBCODE, BKTYPE, BKPRICE, PB)
     Primary key BKCODE
     Foreign key PUBCODE matches PUBLSHR
     1.    BKCODE must be unique.
     2.    PUBCODE must match the code of a publisher in the
           PUBLSHR table.

WROTE(BKCODE, AUTHNUMB, SEQNUMB)
     Primary key BKCODE, AUTHNUMB
     Foreign key BKCODE matches BOOK
     Foreign key AUTHNUMB matches AUTHOR
     1.    The combination of BKCODE and AUTHNUMB must be unique.
     2.    BKCODE must match the code of a book in the
           BOOK table.
     3.    AUTHNUMB must match the number of an author in the
           AUTHOR table.

INVENT (BKCODE, BRNUMB, OH)
     Primary key BRNUMB, BKCODE
     Foreign key BRNUMB matches BRANCH
     Foreign key BKCODE matches BOOK
     1.    The combination of BKCODE and BRNUMB must be unique.
     2.    BRNUMB must match the number of a branch in the
           BRANCH table.
     3.    BKCODE must match the code of a book in the
           BOOK table.
```

CHAPTER 7 — FUNCTIONS OF A DATABASE MANAGEMENT · SYSTEM

1. The DBMS must furnish a mechanism for storing data and for enabling users to retrieve data from the database and to update data in the database. This mechanism should not require the users to be aware of the details with respect to how the data is actually stored.

3. Shared update refers to two or more users updating data in a database at the same time.

5. Locking is the process whereby only one user is allowed to access a specific portion of a database at a time. When a user is accessing a portion of the database, it is locked, meaning that it is unavailable to any other user.

7. Deadlock is the circumstance in which user A is waiting for resources that have been locked by user B, and user B is waiting for resources that have been locked by user A; unless action to the contrary is taken, the two will wait for each other forever. It occurs when each of the two users is attempting to access data that is held by the other.

9. a. Each user must attempt to lock all the resources he or she needs before beginning any updates. If any of the resources are already locked by another user, all locks must be released and the process must begin all over again.

   b. Before updating a record, user 1 should make sure that the record has not been updated by user 2 since the time user 1 first read it. To understand why this is necessary, read part c of this answer.

    c. After reading a record, a user should immediately release the lock on it.

11. Security is the prevention of unauthorized access to the database.

13. Encryption is the process whereby data is transformed into another form before it is stored in the database. The data is returned to its original form when it is retrieved by a legitimate user. This process prevents a person who bypasses the DBMS and accesses the database directly from seeing the relevant data.

15. A database has integrity when the data in it follows certain established rules, called integrity constraints. Integrity constraints can be handled in four ways: (1) They can be ignored. (2) The responsibility for enforcing them can be assigned to the user (that is, it would be up to the user not to enter invalid data). (3) They can be enforced by programs. (4) They can be enforced by the DBMS. Of these four, the most desirable is the last. When the DBMS enforces the integrity constraints, users don't have to constantly guard against entering incorrect data, and programmers are spared having to build the logic to enforce these constraints into the programs they write.

17. Many examples are possible; if you need help in remembering some, see the list in this chapter.

CHAPTER 8 — DATABASE ADMINISTRATION

1. DBA is database administration, the person or group that is responsible for the database. The responsibilities of DBA are crucial to success in the database environment, especially if the database is to be shared among several users; these responsibilities include determining access privileges; establishing and enforcing security procedures; determining and enforcing policies with respect to the use of a data dictionary; and so on.

3. DBA determines access privileges, uses the DBMS security facilities such as passwords, encryption, and views, and supplements these features, where necessary, with special programs.

    Users often choose passwords that are easy for others to guess, such as the names of family members. Users can also be careless with the paper on which passwords are written. To prevent others from guessing their passwords, users should guard against doing either of these things and should also change their passwords frequently.

5. Certain corporate data, though no longer required in the active database, must be kept for future reference. A data archive is a place for storing this type of data. The use of data archives allows an organization to keep records indefinitely, without causing the database to become unnecessarily large. Data can be removed from the database and placed in the data archive, instead of just being deleted.

7. DBA does some of the training of computer users. Other training, such as that which is provided by a software vendor, is coordinated by DBA.

9. a. What facilities are provided by the system for defining a new database? What data types are supported?

    b. What facilities are present to assist in the restructuring of a database?

    c. What nonprocedural language (a language in which we tell the computer what the task is rather than how to do it) is furnished by the system? How does its functionality compare with that of SQL?

    d. What procedural language (a language in which we tell the computer how to do the task) is provided by the system? How complete is it? How is the integration between the procedural language and the nonprocedural language accomplished?

    e. What data dictionary is included? What types of information can be held in the dictionary? How well is it integrated with the other parts of the system?

f. What support does the system provide for shared update? What type of locking is used? Can the system handle deadlock?

g. What services does the system provide for backup and recovery? Does recovery consist only of copying a backup over the live database, or does the system support the use of a journal in the recovery process?

h. What security features are provided by the system? Does it support passwords, encryption, and/or views? How easy is it for a user to bypass the security controls of the DBMS?

i. What type of integrity support is present? What kinds of integrity constraints can be enforced?

j. What are the system limitations with respect to the number of tables, columns, rows, and the number of files that can be open at the same time? What hardware limitations exist?

k. How good are the manuals? How good is the on-line help facility, if there is one?

l. What reputation does the vendor have for support of their products?

m. How well does the system perform?

n. What is the cost of the DBMS, of additional hardware, and of support?

o. What plans does the vendor have for further development of the system?

p. This category includes any special requirements an organization might have that do not fit into any of the previous categories.

11. DBA has primary responsibility for the DBMS once it has been selected. DBA installs the DBMS, makes any changes to its configuration when they are required, determines whether it is appropriate to install a new version of the DBMS when it becomes available, and, if a decision is made to install a new DBMS, coordinates the installation.

CHAPTER 9 — APPLICATION GENERATION

1. An application system is a collection of programs that is designed to handle some specific task. Application systems can be created by users, by programmers within the organization that will use the system, and by software houses. A DBMS is designed to handle general rather than specific tasks and is thus not an application system. Many DBMSs are used to write application systems, however.

3. A modern application system is menu-driven and features interactive and automatic updates, reports, some type of query facility, backup and recovery facilities, audit trails, utility services, and administrative services.

5. Interactive updates are usually achieved through users filling in forms on the screen. Not all updates are accomplished interactively, however. Some occur in response to some other update (for example, a customer's balance is increased automatically as a result of printing an invoice) or in response to a particular request from a user (for example, month-to-date fields are set back to zero when the user chooses the month-end routines).

7. An audit trail is a record of all the updates that have been made to the database. It is used to ensure that updates have been made correctly and, if any errors have been found, to facilitate restoring the database to a correct state.

9. A programmer's workbench consists of a collection of tools, such as editors, compilers, and debuggers, that assist the programmer in doing his or her job.

11. A screen painter is a tool that is used to facilitate the process of developing screen-oriented data entry programs. To use it, a programmer interactively describes what the screen is supposed to look like. A screen painter is also called a screen generator.

13. A nonprocedural language is one in which a user describes what the task is rather than how it is to be accomplished. SQL is an example of a nonprocedural language. Nonprocedural languages make users more productive, because they do not need to be concerned with specific details with regard to how a task is to be accomplished.

15. An application generator needs a nonprocedural language because users are more productive if the task alone is specified, not how to do it. An application generator also needs a procedural language to handle any activities that are beyond the capabilities of the nonprocedural language. To obtain the maximum benefit from both languages, it should be possible to embed commands from the nonprocedural language in the middle of procedural language programs.

17. A help facility is a feature through which users can receive on-line assistance for the task in which they are currently engaged. If we're creating an application system for a group of users and we're using an application generator that does not include a help facility in the application systems we develop, we must include the logic to provide this help in our programs.

19. Describe the contents of the database to the data dictionary. Use the screen painter to create forms for interactive updates. Enter sample data. Create reports or queries using the report and query facilities of the application generator. Use the nonprocedural language to add any missing features. Then, if any features are still missing, use the procedural language. Tie all of the above together in a menu with the menu generator. Include a help facility.

Glossary

Alias An alternate name for a table; can be used within a query.

ANSI American National Standards Institute.

Application See *application system*.

Application generation The process of developing an application system.

Application generator Software tool that permits the rapid development of an application system. Also called a *fourth-generation environment* or a *fourth-generation language*. The latter term is sometimes applied only to the *nonprocedural language* component of an application generator.

Application package See *application system*.

Application programs The programs that make up an *application system*.

Application system A collection of programs that work together to handle some specific task. Also called an *application*, an *application package*, a *software package*, a *software system*, or sometimes simply a *system*.

Attribute A property of an entity.

Background The permanent part of a screen form, that is, the part that does not change from one transaction to the next. See also *foreground*.

Backup A copy of a database; used to recover the database when it has been damaged or destroyed.

Boyce-Codd Normal Form (BCNF) A relation is in Boyce-Codd normal form if it is in second normal form and the only determinants it contains are candidate keys; also called third normal form in this text.

Candidate key A minimal collection of attributes (columns) in a relation on which all attributes are functionally dependent but which has not necessarily been chosen as the *primary key*.

Catalog A source of information on the types of entities, attributes, and relationships in a database.

Child In the hierarchical model, the record that is the "many" part of the *one-to-many relationship*.

CODASYL COnference on Data SYstems Languages. The group that developed COBOL and proposed the CODASYL model for database management.

CODASYL model The model for database management systems proposed by CODASYL; falls within the general network model of data. Not a standard, although it has been used in the development of many systems.

Concatenation Combination of attributes. To say a key is a concatenation of two attributes, for example, means that a combination of values of both attributes is required to uniquely identify a given tuple.

Concurrent update Several updates taking place to the same file or database at almost the same time; also called *shared update*.

Data archive A place where historical corporate data is kept. Data that is no longer needed in the corporate database but must be retained for future needs is removed from the database and placed in the archive.

Data Base Task Group The group originally appointed by CODASYL to develop specifications for database management systems.

Data definition language A language that is used to communicate the structure of a database to the database management system.

Data dictionary A tool that is used to store descriptions of the entities, attributes, relationships, programs, and so on, that are associated with an organization's database.

Data independence The property that allows the structure of the database to change without requiring changes in programs.

Data manipulation language A language that is used to manipulate the data in the database.

Data model A classification scheme for database management systems. A data model addresses two aspects of database management: *structure* and *operations*.

Data structure diagram A diagram of the records and sets in a network database.

Database A structure that can house information about various types of entities and about the relationships among the entities.

Database administration (DBA) The individual or group that is responsible for the database.

Database administrator The individual who is responsible for the database, or the head of database administration.

Database design The process of determining the content and arrangement of data in the database in order to support some activity on behalf of a user or group of users.

Database Design Language (DBDL) A relational-like language that is used to represent the result of the database design process.

Database management system A software package that is designed to manipulate the data in a database on behalf of a user.

Database navigation The process of finding a path through the relationships in a database in order to satisfy a given request.

Database processing The type of processing in which the data is stored in a *database* and manipulated by a *DBMS*.

DBA See *database administration*. (Sometimes the acronym stands for database administrator.)

DBDL See *database design language*.

DBMS See *database management system*.

DDL See *data definition language*.

Deadlock A state in which two or more users are each waiting to use resources that are held by the other(s).

Deadly embrace Another name for *deadlock*.

Debugging The process of finding and correcting errors in programs.

Defining query The query that is used to define the structure of a view.

Dependency diagram A diagram that indicates the dependencies among the attributes in a relation.

Determinant An attribute that determines at least one other attribute.

DML See *data manipulation language*.

Encryption The transformation of data into another form, for the purpose of security, before it is stored in the database. The data is returned to its original form for any legitimate user who accesses the database.

Entity An object (person, place, or thing) of interest.

Entity integrity The rule that no attribute that participates in the primary key may accept **null** values.

Field The smallest unit of data to which we assign a name; can be thought of as the columns in a table. For example, in a table for customers, the fields (columns) would include such things as the customer's number, the customer's name and address, and so on.

File Technically, a collection of bytes (characters) on a disk; could be data, a program, a document created by a word processor, and so on. Often refers to a data file, which is a structure used to store data about some entity. Such a file can be thought of as a table. The rows in such a table are called records, and the columns are called fields.

First Normal Form (1NF) A relation is in first normal form if it does not contain repeating groups. (Technically, this is part of the definition of a relation.)

Foreground The portion of a screen form that changes from one transaction to the next, that is, the portion of the form into which the user enters data and/or data is displayed. See also *background*.

Foreign key An attribute (or collection of attributes) in a relation whose value is required either to match the value of a primary key in another relation or to be null.

Fourth-generation environment See *application generator*.

Fourth-Generation Language (4GL) Sometimes used to refer to an *application generator* and sometimes to a *nonprocedural language*, which is one of the components of an *application generator*.

Fully relational The expression used to refer to a DBMS in which users perceive data as tables, which supports all the operations of the *relational algebra* and which supports *entity* and *referential integrity*.

Functionally dependent Attribute B is functionally dependent on attribute A (or on a collection of attributes) if a value for A determines a single value for B at any one time.

Functionally determine Attribute A functionally determines attribute B if B is *functionally dependent* on A.

Help facility A facility through which users can receive on-line assistance.

Hierarchical model A data model in which the structure is a *tree*, or hierarchy.

Hierarchy See *tree*.

Index A file that relates key values to records that contain those key values.

Information level of database design The step during *database design* in which the goal is to create a clean, DBMS-independent design that will support user requirements.

Integrity A database has integrity if all *integrity constraints* that have been established for it are currently met.

Integrity constraint A condition that data within a database must satisfy; also, a condition that indicates the types of processing that may or may not take place.

Join In the *relational algebra*, the operation in which two tables are connected on the basis of common data.

Journal A record of all changes in the database; also called a *log*. Used to recover a database that has been damaged or destroyed.

LAN See *local area network*.

Local Area Network (LAN) A configuration of several computers that are all hooked together in a limited geographic area; allows users to share a variety of resources.

Locking The process of placing a lock on a portion of a database, which prevents other users from accessing that portion.

Log A record of all changes in the database; also called a *journal*. Used to recover a database that has been damaged or destroyed.

Logical child relationship A relationship in the *hierarchical model* in which the *parent* is in a different *tree* than the *child*.

Many-to-many relationship A relationship between two entities in which each occurrence of each entity is related to many occurrences of the other entity.

Mapping The process of creating an initial *physical-level design*.

Member In a *CODASYL set*, the record that is the "many" part of the *one-to-many relationship*.

Menu-driven A style of program in which the user selects an action from a list (called a menu) of available options that are displayed on the screen.

Minimally relational A DBMS in which users perceive data as tables and which supports at least the SELECT, PROJECT, and JOIN operations of the *relational algebra* without requiring any predefined access paths.

Natural language A language in which users communicate with the computer through the use of standard English questions and commands.

Navigation See *database navigation*.

Network A structure that contains record types and explicit one-to-many relationships between these record types.

Network model A *data model* in which the structure is a network and the operations involve navigating the network (that is, following the arrows in a data structure diagram).

Nonkey attribute An attribute that is not part of the primary key.

Nonprocedural language A language in which the user specifies the task that is to be accomplished rather than the steps that are required to accomplish it.

Normal form See *first normal form*, *second normal form*, *third normal form*, and *Boyce-Codd Normal Form*.

Normalization Technically, the process of removing repeating groups to produce a *first normal form* relation. Sometimes refers to the process of creating a *third normal form* relation.

Null A special value meaning "unknown" or "not applicable."

One-to-many relationship A relationship between two entities in which each occurrence of the first entity is related to many occurrences of the second entity but each occurrence of the second entity is related to only one occurrence of the first entity.

One-to-one relationship A relationship between two entities in which each occurrence of the first entity is related to one occurrence of the second entity and each occurrence of the second entity is related to one occurrence of the first entity.

Operations One of the two components of a *data model*; the facilities given users of the DBMS to manipulate data within the database.

Optimizer The DBMS component that selects the best way to satisfy a query.

Owner In a CODASYL set, the record that is the "one" part of the one-to-many relationship.

Parent In the hierarchical model, the record that is the "one" part of the one-to-many relationship.

Partial dependency A dependency of an attribute on only a portion of the primary key.

Password A word that must be entered before a user can access certain computer resources.

Physical database In the *hierarchical model*, a single *tree*.

Physical level of database design The step during *database design* in which a design for a given DBMS is produced from the final information-level design.

Primary key A minimal collection of attributes (columns) in a relation on which all attributes are functionally dependent and which is chosen as the main direct-access vehicle to individual tuples (rows). See also *candidate key*.

Procedural language A language in which the user must specify the steps that are required for accomplishing a task instead of merely specifying the task itself.

Program generator A facility that accepts specifications from the user and creates a program that matches these specifications.

Programmer's workbench A collection of tools that simplifies the tasks of programmers; includes editors, compilers, debuggers, and so on.

QBE See *Query-by-Example*.

Qualify To indicate the table (relation) of which a given column (attribute) is a part by preceding the column name with the table name. For example, *CUSTOMER.ADDRESS* indicates the column named *ADDRESS* within the table named *CUSTOMER*.

Query A question, the answer to which is found in the database; also used to refer to a command in a *nonprocedural language* such as SQL that is used to obtain the answer to such a question.

Query-By-Example (QBE) A *data manipulation language* for relational databases in which users indicate the action to be taken by filling in portions of blank tables on the screen.

Query facility A facility that enables users to obtain information easily from the database.

Query language A language that is designed to permit users to obtain information easily from the database.

Record A collection of related fields; can be thought of as a row in a table.

Recovery The process of restoring a database that has been damaged or destroyed.

Redundancy Duplication of data.

Referential integrity The rule that if a relation A contains a *foreign key* that matches the primary key of another relation B, then the value of this foreign key must either match the value of the primary key for some row in relation B or be null.

Relation A two-dimensional table in which all entries are single-valued; each column has a distinct name; all of the values in a column are values of the attribute that is identified by the column name, the order of columns is immaterial; each row is distinct; and the order of rows is immaterial.

Relational algebra A relational **data manipulation language** in which relations are created from existing relations through the use of a set of operations.

Relational database A collection of relations.

Relational model A *data model* in which the structure is the *table* or *relation*.

Relationally complete A term applied to any relational *data manipulation language* that can do whatever can be done through the use of the *relational algebra*; also applied to a DBMS that supplies such a data manipulation language.

Relationship An association between entities.

Repeating group Several entries at a single location in a table.

Report generator See *report writer*.

Report writer A nonprocedural language for producing formatted reports from data in a database; also called a *report generator*.

Save A backup copy.

Screen generator An interactive facility for creating and maintaining display and data-entry formats for screen forms.

Screen painter See *screen generator*.

Second Normal Form (2NF) A relation is in second normal form if it is in first normal form and no non-key attribute is dependent on only a portion of the primary key.

Secondary key An attribute or collection of attributes that is of interest for retrieval purposes (and that is not already designated as some other type of key).

Security The protection of the database against unauthorized access.

Set The CODASYL implementation of a one-to-many relationship.

Shared update Several updates taking place to the same file or database at almost the same time; also called *concurrent update*.

Software package See *application system*.

Software system See *application system*.

SQL See *Structured Query Language*.

Structure One of the two components of a *data model*: the manner in which the system structures data or, at least, the manner in which the users perceive that the data is structured.

Structured Query Language (SQL) A very popular relational *data definition and manipulation language* that is used in many relational DBMSs.

Subquery In SQL, a query that is contained within another query.

System Often used to refer to *application system*; sometimes also used to refer to a *DBMS*.

Table In the database environment, another name for a relation.

Tabular A type of DBMS in which users perceive data as tables but which does not furnish any of the other characteristics of a relational DBMS.

Third Normal Form (3NF) A relation is in third normal form if it is in second normal form and if the only determinants it contains are candidate keys. (Technically, this is the definition of *Boyce-Codd Normal Form*, but in this text the two are used synonymously.)

Tree A network, with an added restriction: no entity can participate as the "many" part of more than one *one-to-many relationship*.

Tuning The process of altering a database design in order to improve performance.

Tuple The formal name for a row in a table.

Unnormalized relation A structure that satisfies the properties required to be a relation with one exception: repeating groups are allowed; that is, the entries in the table do not have to be single-valued.

Update anomaly An update problem that can occur in a database as a result of a faulty design.

User view The view of data that is necessary to support the operations of a particular user.

View An application program's or an individual user's picture of the database.

Wild card A symbol that can be used in place of an unknown character or group of characters in a query.

Bibliography

[1] Boar, Bernard H. *Application Prototyping: A Requirements Definition Strategy for the 80s*. John Wiley & Sons, 1984.

[2] Chen, Peter. *The Entity-Relationship Approach to Logical Data Base Design*. QED Monograph Series, 1977.

[3] Codd, E. F. "A Relational Model of Data for Large Shared Databanks." *Communications of the ACM* 13, no. 6 (June 1970).

[4] Codd, E. F. "Further Normalization of the Data Base Relational Model." In *Data Base Systems*, Courant Computer Science Symposia Series, vol. 6, Prentice-Hall, 1972.

[5] Codd, E. F. "Recent Investigations into Relational Data Base Systems." Proceedings of the IFIP Congress, 1974.

[6] Codd, E. F. "Extending the Relational Database Model to Capture More Meaning." *ACM TODS* 4, no. 4 (December 1979).

[7] Codd, E. F. "Relational Database: A Practical Foundation for Productivity." *Communications of the ACM* 25, no. 2 (February 1982).

[8] Date, C. J. *Database: A Primer*. Addison-Wesley, 1983.

[9] Date, C. J. *Introduction to Database Systems: Volume I*, 4th ed. Addison-Wesley, 1986.

[10] Goldstein, Robert C. *Database Technology and Management*. John Wiley & Sons, 1985.

[11] Kroenke, David. *Database Processing*, 2d ed. SRA, 1983.

[12] Kroenke, David M., and Nilson, Donald E. *Database Processing for Microcomputers*. SRA, 1986.

[13] McFadden, Fred R., and Hoffer, Jeffrey A. *Data Base Management*. Benjamin Cummings, 1985.

[14] Pratt, Philip J., and Adamski, Joseph J. *Database Systems: Management and Design*, 3rd ed. boyd & fraser, 1994.

[15] Vasta, Joseph A. *Understanding Data Base Management Systems*. Wadsworth, 1985.

[16] Zloof, M. M. "Query By Example." Proceedings of the NCC 44, May 1975.

Abort, 136
  during update, 123
Access, 15, 16
  database administration and,
    133
  locked, 119-123
  restricting, 124-125
  unauthorized, 134
Additions, update anomalies
    and, 80, 83
Administrative services, appli-
    cation system and, 149
Algorithm, 154
Alias, 57
Alphanumeric field, 41
Alternate key, 78, 94
  database design and, 94
  database design language
    and, 95
Alternate names, natural lan-
    guages and, 57
Alter table, 70
AND
  QBE and, 53
  SQL and, 43
Application(s), 144
Application generation,
    144-157
  components of, 149-155
  data dictionary and, 150
  introduction to, 144-145
  features of, 149
  help facility and, 154-155
  menu generator and, 154-155
  nonprocedural language and,
    154
  other terminology, 157
  procedural language and, 154
  program generator and, 155
  programmer's workbench
    and, 150
  query facility, 153
  relational DBMS and, 150
  report writer, 153-154
  screen painter, 151
  utilization and, 155-156
Application package, 3

Application programs, 3
  integrity constraints and, 125
Applications generator, 150
Application system, 3, 144
  administrative services and,
    149
  audit trails and, 149
  backup and, 149
  help facility and, 154-155
  interactive and automatic
    updates and, 146-148
  queries and, 148
  recovery and, 149
  reports and, 149
  utility services and, 149
Archives, database admini-
    stration and, 137
Ashton-Tate, 13
Assembly language, 157
Asterisk (*)
  database design language
    and, 95
  SQL and, 42, 45, 128
  views and, 63
Attribute, 10, 26
  candidate key and, 78
  determinant, 83
  functionally dependent, 75-77
  primary key and, 77-78
  nonkey, 80
  relational model and, 25
  reordering, 97
Audit trails, application sys-
    tem and, 149
AVG, SQL and, 45

Background, 151
Backup, 113, 123
  application system and, 149
  database administration and,
    136
  DBMS evaluation and, 140
BASIC, 3, 8, 149, 154, 157
Benchmark tests, 141
Blank value, 68
Blinking, 152
Bookstores case, 2-17
Boxes, drawing on screen, 152

Boyce-Codd normal form, **74**
Built-in-functions, SQL and,
    45-46

Candidate key, **78, 83**
  ensuring uniqueness and, 108
Catalog, **67, 113**
  relational model and, 67-68
  SQL and, 67-68
CHANGE COLUMN NAME
    TO, 70
CHAR, SQL and, 41
Character field, 152
  SQL and, 41
Checklist, evaluation and
    selection of DBMS and,
    138-139
Children, **34**
COBOL, 8, 154, 157
CODASYL model, **30**
Codd, E.F., 68, 71, 74
College courses case, 2
Colors, screen and, 152
COLTYPE, 67
Column(s),
  addition of, 128
  changing length of, 128
  deleting, 70
  determinant, 83
  indexes for, 65
  naming, 43, 63
  null value, 68
  renaming with views, 63
  SQL and, 41
Column heading
  QBE and, 52
  SQL and, 43
Commas, SQL and, 40
Comparison operators
  QBE and, 53
  SQL and, 42, 43
Compilation, 150
Compound conditions
  QBE and, 53
  SQL and, 43, 47, 48
  update and, 50-51
  WHERE clause and, 43
Concatenation, 56, **76**

Constraints, enforced by pro-
    grams, 108
Copying files, 123
Costs, DBMS and, 15
COUNT, 45
CREATE TABLE, 51
CREATE VIEW, 63
Credit, available, 43
Customer information, 21
Customer name, deletion and,
    50
Customer numbers, 50, 65

Data
  consistency and, 16
  damaged, 123
  duplication of, 13
  inconsistent, 80, 82
  redundant, 16
  sample, 156
  sharing of, 15
Data archive, **137**
Database, **3, 12**
  damaged, 123
  integrity, 16
  pictorial representation and,
    96
  writing programs to
    maintain, 3-6
Database administration
    (DBA), **16, 132-142**, 149
  access privileges and, 133
  archives and, 137
  backup and, 136
  database design and, 142
  data dictionary management,
    138
  DBMS evaluation and selec-
    tion and, 138
  DBMS support and, 138-142
  introduction to, 132-133
  naming conventions and, 138
  passwords and, 135
  policy formulation and
    implementation
    and, 133-137
  recovery and, 136
  training and, 138
Database approach, **14, 15**

Database design, **8, 91**
  database administration and,
    142
  documentation of, 142
  examples, 97-107
  general methodology, 92
  identifying keys, 94-95
  information-level, 92-107
  introduction to, 91
  merging result into, 96-97
  methodology for, 91-109
  normalization, 74, 89, 94
  physical-level, 91, 108-109
  pictorial representation and,
    96
  representing user view as col-
    lection of tables, 92-94
  tuning, 142
  user views, 92-102
Database design language
  (DBDL), 95-96
Database processing, **13**
  advantages of, 15-17
  disadvantages of, 17-18
  hierarchical model and, 34-35
  network system and, 30-34
  shorthand representation of,
    26-27
  utility services and, 129
Data definition
  database administration and,
    138
  DBMS evaluation and, 140
  SQL and, 41
Data dictionary, **115, 138, 150**
  application generator and,
    140
  DBMS evaluation and, 140
  screen generator and, 152
Data file, **4**, 12
Data independence, **17, 113,**
  127-129
  relational model and, 30
Data manipulation, 39-59
  natural languages and, 57-59
  QBE and, 51-55
  relational algebra and, 55-57
  SQL and, 40-51
Data model, **20**
  hierarchical model and, 34-36

network model and, 30-34
  Premiere Products and, 21-24
  relational model and, 25-30
Data restructuring, DBMS
  evaluation and, 140
Data Retrieval, **113**
  indexes and, 64-66
Data storage, **113**
  encrypted format, 124
Data type, 125
Data-type constraints,
    microcomputer DBMSs
    and, 127
Data validation rules,
    database administration
    and, 138
Date field, 152
dBASE, 12
DBMS (database management
    system), **8, 12-13**
  background, 10-14
  evaluation and selection, 138-
    142
  introduction to, 1-18
  programming with, 8-9
  responsibility for, 141
  software packages and, 3-6
  terms, 10-14
DBMS functions, 113-129
  catalog, 113
  data independence, 127-129
  data retrieval, 113
  data storage, 113
  integrity, 125-127
  recovery, 123, 124
  security, 124-125
  shared update, 115-123
  support, 138-142
  utilities, 129
Deadlocks, 120-121
Debugging, 150
DECIMAL, SQL and, 41
Decimal places, 152
Defining query, **62**, 128
Delete/deleting
  column, 70
  update anomalies and, 80, 83
DELETE, SQL and, 50
Dental practice case, 2, 10
Dependencies, partial, **80**

Dependency diagram, 80, 83
Determinant, 83
Diskette archives, 137
Disk problem, 136
Documentation
  database design and, 141
  DBMS evaluation and, 141
DOS services, 129
DROP TABLE, 70

Edit capabilities, 129
Efficiency
  developing application sys-
      tems and, 150
  indexes and, 64
  network model and, 36
  relational model and, 30
Encryption, 124, 149
Entities, 3, 10
  database design and, 93
  determining properties for, 93
  relationships between, 3,
      93-94
Entity integrity, 68
Equality, SQL and, 43
Error correction, utility serv-
      ices and, 149
Error message, 29, 147
Export, 129

Failure, 18
Fields, 30,
Files, 3, 12
  copying, 123
FIND NEXT WITHIN
      PLACED, 33
FIND NEXT WITHIN
      REPRESENTS, 32
FIND OWNER WITHIN
      REPRESENTS, 32, 34
First normal form (1NF), 74,
      78, 87
  normalization and, 78-79
  update anomalies and, 80
Foreground, 152
Foreign key, 69, 94
  constraints, 125
  database design and, 94, 95
  database design language
      and, 96

legitimate values, 108
  one-to-many relationship
      and, 94
Form(s), screen painting and,
      151-153
Format
  integrity constraints and, 125
  keys and, 125
Format constraints, 127
Formatting, reports and,
      153-154
FORTRAN, 8, 154, 157
Fourth-generation environ-
      ment, 129, 144, 157
Fourth-generation language
      (4GL) 144, 157
Fully relational, 72
Functional dependence, 80
  normalization and, 75-77
  split across two table, 85
Functionally dependent, 76
Functionally determines, 76
Future plans, DBMS evalua-
      tion and, 141

GIVING, 55
Greater than (>), 43
Greater than or equal to
      (> =), 43

Hardware problem 123, 136
Help, context sensitive, 141
Help facility, 154-155
Hierarchical model, 34-36
Hierarchy, 34
High-level languages, 157
Highlight, 152

IBM (International Business
      Machines), 28
Import, 129
Index(es), 64, 109, 150
  creating, 128
  deletion of, 129
  naming conventions and, 138
  opening database and, 128
  relational model and, 64-66
  updates and, 66
  utility services and, 129

Information, categories of, 3
Information-level design,
  91-107
  DBA and, 141
  examples, 97-107
  general methodology, 92
  identifying keys, 94-95
  merging result into design
    and, 96-97
  normalize, 94
  pictorial representation and,
    96
  representing user view as col-
    lection of tables, 92-94
  user views and, 92-102
INSERT, SQL and, 50, 51
Integrity, 30, 113
  DBMS evaluation and, 140
  entity, 68
  referential, 68-69
Integrity constraints, 16, 30,
  125-127
Integrity rules, 68-69, 72
Interface, menu-driven, 129

Joining/joins, 54, 56
  multiple tables, 48-50
  QBE and, 54-55
  relational algebra and, 56-57,
    71
  restricting rows in, 47
  SQL and, 47
  two tables, 46-47
  view using, 63
Journal, 123

Key(s)
  data type and, 125
  identifying in database
    design, 94
  legal values and, 125
  normalization and, 77-78
  physical-level design and, 108
  sorting on multiple, 45
Key constraints, 127

Languages, 8
  machine, 157
  natural, 57

nonprocedural, 129, 140, 154
procedural, 129, 140, 154
SQL, 40-51
utility services and, 129
Legal values, 125, 126
Less than (<), 43
Line items, 21
Local area networks (LANs),
  140
Locked record/locking,
  119-123
  deadlock and, 120-121
  duration of, 120
  microcomputer DBMSs and,
    121-122
Logical child relationship, 35

Machine languages, 157
Mainframe DBMSs, 13, 15
  archives and, 137
  indexes and, 66
  locking and, 121
  recovery and, 123
  relationally complete, 72
Many-to-many relationship, 94
MAX, 45
Member, 31
Menu(s)
  application system, 145-146
  creating using applications
    generator, 154
Menu-driven, 145, 156
Microcomputer DBMSs, 13
  archives and, 137
  catalog and, 115
  costs, 15
  features, 113
  indexes and, 126-127
  integrity and, 126-127
  locking and, 119-123
  physical-level design and, 108
  recovery and, 123-124, 136
  relationally complete, 72
Microcomputer file-
  management systems, 72
Microrim, 13
MIN, 45
Minimally relational, 72

NAME column, 67
Naming
  columns, 43, 63
  DBA conventions, 138
  indexes, 138
  tables, 138
Natural languages, 57-59
Nesting queries, 47-48
Network(s) **30**
  local area, 140
  update and, 115
Network model, **30**
  advantages and disadvantages
    of, 35-36
  operations within, 31-33
  structure within, 30-31
Nonkey attribute, **80**
Nonprocedural language, 129,
  140, 154
Normal forms, **74**
Normalization, **74**
  database design and, 94
  first normal form, 78-79
  functional dependence and,
    75-77
  incorrect decompositions, 85-
    87
  information-level design, 94
  introduction to, 74-75
  keys and, 77-78
  second normal form, 79-82
  third normal form, **82-84**
NOT, 43, 53-54
Not equal to (>), 43
Null, **68**
  altered tables accepting, 70
  database design language
    and, 96
  foreign keys and, 68
Numeric field, screen painter
  and, 152

One-to-many relationship
  database design and, 93
  hierarchical model and,
    34, 35
  network model and, 31
  relational model and, 25
One-to-one relationship, 94

On-line help, 141
Optimizer, **66**
OR, 43, 53
ORDER BY, SQL and, 40, 44-
  45
Order lines, 21-23
Order number, 21
Order total, 21
Owner, **31**

Parentheses(), SQL subqueries
  and, 47-48
Parents, **34**
Part description, 21
Partial dependencies, **80**
Pascal, **8**, 154, 157
Password, 124, 135, 149
Payroll system, 144
Percent symbol (%), SQL and,
  44
Performance, DBMS evalua-
  tion and, 141
Physical database, 35
Physical-level design, 91,
  108-109, 142
Policy, database admini-
  stration and, 133-137
Power failure, 136
PREMIERE PRODUCTS appli-
  cation, 21-24
Primary key, **28, 68, 77, 83**
  constraints, 125
  database design and, 93
  ensuring uniqueness and, 108
  more than one possibility for,
    78
Printed archives, 137
Printer management, 9
Printing, QBE and, 52-55
Procedural language, 129
  application generators and,
    154
  DBMS evaluation and, 140
Procedure, 154
Productivity, 17
Program generator, **155**
Programmers, 17, 144
  catalog and, 113

enforcing restrictions, 108
integrity constraints and, 125
productivity of, 157
Programmer's workbench, 150
Programming, 8-9
PROJECT, relational algebra
 and, 56, 57, 71

QBE, see Query-By-Example
Qualify, 46
Qualify data, 27
Queries, 129, 148
 ad hoc, 153
 application system and, 148
 catalog and, 67-68
 defining, 128-129
 natural languages, 57
 nesting, 47-48
 relational algebra and, 55-57
 SQL and, 40-51
 two tables, 46-47
Query-By-Example (QBE),
 51-55, 153
Query facility, 153, 156

R:BASE, 13
Record, 4
Record number, 65
Record types, 30
Recovery, 18, 113, 123-124
 application system and, 149
 database administration and,
  136
 DBMS evaluation and, 140
Redundancy, 16, 80, 85
Referential integrity, 68, 69, 95
 database design and, 96
 lack of support for, 72
Relation, 25, 26
 unnormalized, 26, 78
Relational, definitions of, 72
Relational algebra, 55-57, 71
 JOIN and, 56-57
 PROJECT and, 55-56, 57
 SELECT and, 55-56
Relational database, 26
 changing structure of, 69-70
 shorthand representation of
  structure, 26

Relationally complete, 72
Relational model, 25-30
 advanced features, 61-72
 advantages and disadvan-
  tages, 29-30
 applications generators and,
  150
 catalog and, 67-68
 data definition and manipulation and, 39-59
 data independence and, 30
 efficiency and, 30
 indexes and, 64-66
 integrity rules and, 68-69
 operations within, 28-29
 views and, 61-64
Relational system, defined, 71
Relationships, 11
 adding, 128
 changes in, 128
 determining in database
  design, 93-94
 explicit, 30, 68
 implicit, 30
 many-to-one, 94
 one-to-one, 94
 relational model and, 26
Repeating groups, 26, 78
Report(s)
 application system, 148
 formatting and, 153-154
Report generators, 129
Report writer, 153, 155, 156
Rows, concatenation of, 56

Save/saving, 123
Screen, color, 152
Screen generator, 129, 151
Screen management, 9
Screen painter, 151-153, 156
Secondary key, 94
 database design and, 94
 database design language
  and, 96
 efficiency of access to rows
  and, 108
Second normal form (2NF),
 74, 80, 87
 normalization and, 79-82
 update anomalies and, 80

Security, 16, **113**, **124**, **134**
  database administration and,
    134-136
  DBMS evaluation and, 140
  encryption and, 124
  passwords and, 124
  views and, 61, 125
SELECT, 55-56, 71
SELECT...FROM...WHERE,
  28, 41-44
Service-oriented organization
  case, 1
Set, **31**
Software packages, **3**
Software system, **3**
Software version, releases of,
  142
Sorting, 44-45
Special effects, screen and, 152
SQL (Structured Query Language) **28**, **153**, **154**
  advanced relational topics
    and, 61-72
  altering tables, 70
  built-in functions and, 45-46
  catalog and, 67-68
  changing database structure
    and, 69-70
  column headings and, 43
  command form, 28
  compound conditions and, 43
  data definition and, 41
  deleting columns, 70
  dropping table and, 70
  querying two tables, 46-47
  simple retrieval and, 41-44
  sorting and, 44-45
  subqueries and, 47-48
  update and, 50-51
  views and, 61-64
Structure, 17
Structured Query Language,
  see SQL
Submenus, 146
Subquery, 47-48
SUM, SQL and, 45
SYSCOLUMNS, 67
SYSTABLES, 67
System, **3**
System of programs, **3**
System parameters, 149

Table(s), **4**, **25**
  altering, 70
  dropping, 70
  incorrect decompositions of,
    85
  joining, 48-50
  naming conventions and, 138
  querying two, 46-47
  representing user views as
    collection of, 92-94
Tabular system, **71**
Tape archives, 137
TBNAME column, 67
Third normal form (3NF), **74**,
  82-84, 87, 97
Totals, 21
Training, database admini-
  stration and, 138
Tree, **34**
Tuning, **142**
Tuple, **26**
Tutorial, 141

Underline, QBE and, 54
Unnormalized relation, **26**, **78**
Update(s), 6, 15, 113
  abort during, 123
  application system and,
    146-148
  automatic, 146, 156
  compound condition and, 50
  indexes and, 66
  interactive, 146-148
  SQL and, 50-51
  views and, 129
Update, shared, **113**, **115-123**
  DBMS evaluation and, 140
  guidelines for, 121-122
  local area networks and, 146
  locking and, 119-123
  problems involved in,
    115-118
Update anomalies, **74**, **80**
  first normal form and, 79
  second normal form and, 80
  third normal form and, 83
User(s)
  integrity constraints and, 125
  training for, 138
User view, **92-102**

Utilities, **113**, 128-129
  application system and, 149
Utility programs, 156

Vendors
  DBMS evaluation and, 141
  training and, 138
View(s), **61**, **125**
  relational model and, 61-64
  row-and-column subset, 63
  security and, 125
  SQL and, 61-64

WHERE clause
  compound condition within,
      43
  listing restrictions in 29
Wild card, SQL and, 44

Zero value, 68
Zloof, M.M., 51